Friederike Krämer
Norbert Mencke

# Flea Biology and Control

The Biology of the Cat Flea
Control and Prevention
with Imidacloprid in Small Animals

# Springer

*Berlin*
*Heidelberg*
*New York*
*Barcelona*
*Hong Kong*
*London*
*Milan*
*Paris*
*Singapore*
*Tokyo*

Friederike Krämer
Norbert Mencke

# Flea Biology and Control

The Biology of the Cat Flea
Control and Prevention
with Imidacloprid in Small Animals

With 49 Figures
and 43 Tables

Springer

DR. FRIEDERIKE KRÄMER
School of Veterinary Medicine Hannover
Department of Parasitology
Bünteweg 17
30559 Hannover
Germany

DR. NORBERT MENCKE
Bayer AG
BG Animal Health
BU Companion Animals
51368 Leverkusen
Germany

Library of Congress Cataloging-in-Publication Data

Krämer , Friederike, 1971-
Flea biology and control: the biology of the cat flea, control and prevention with
imidacloprid in small animals / Friederike Krämer, Norbert Mencke.
p. cm.
    Includes bibliographical references (p.) and index.
    ISBN 3540417761 (alk. paper)
1. Cat flea–Control. 2. Cat flea. 3. Cats–Parasites–Control. 4. Dogs–Parasites–Control.
5. Pets–Parasites–Control. I. Mencke, Norbert, 1959-II. Title.

SF986.C37 K36 2001
628.9'657–dc21                                                    2001020618

ISBN 3-540-41776-1    Springer-Verlag Berlin Heidelberg New York

Springer-Verlag Berlin Heidelberg New York
a member of BertelsmannSpringer Science+Business Media GmbH

http://www.springer.de

© Springer-Verlag Berlin Heidelberg 2001
Printed in Germany

Production: PRO EDIT GmbH, 69126 Heidelberg, Germany
Cover design: design + production GmbH, Heidelberg, Germany
Typesetting and Reproduction of the figures: AM-productions GmbH,
69168 Wiesloch, Germany

Printed on acid-free paper – SPIN 10791514   27/3130/ML – 5 4 3 2 1 0

Dedicated to
all itching and scratching dogs
and cats in the world,
may you not suffer
from flea bites any longer.

# Contents

1  **Introduction** .................................................... 1

2  **General Morphology** ............................................. 3

3  **Taxonomy** ....................................................... 5

4  **Dissemination and Economic,**
   **Veterinary and Medical Importance** ............................. 9
   Origin and Distribution ......................................... 9
   Economic Importance ............................................ 12
   Veterinary Importance .......................................... 13
   Medical Importance ............................................. 14

5  **Developmental Cycle of Fleas** .................................. 17
   Eggs ........................................................... 18
   Larvae ......................................................... 20
   Pupae .......................................................... 25
   Preemerged Adults .............................................. 27
   Adults ......................................................... 29

6  **Flea Epidemiology** ............................................. 35

**7 Flea-Related Dermatitis** .......................................... 39

Flea Bite Dermatitis ................................................ 39

Flea Allergy Dermatitis (FAD) ...................................... 40

Clinical Signs ...................................................... 41
Differential Diagnosis of FAD ....................................... 45
Histopathological Findings .......................................... 46
Immunopathogenesis of FAD ......................................... 47

Diagnosis of FAD and Flea Bite Dermatitis .......................... 51

Treatment of Flea-Related Skin Diseases ............................ 54

Adjunctive Medical Therapy .......................................... 55
Hyposensitization ................................................... 55

Treatment of the Premises .......................................... 56

Indoor Treatment ................................................... 56
Outdoor Treatment .................................................. 58

**8 Resistance** ..................................................... 59

**9 Imidacloprid** ................................................... 63

History of Imidacloprid ............................................. 63

Chemical Properties of Imidacloprid ................................ 64

Biological Profile, Mode of Action
and Direct Insecticidal Activity .................................... 67

Biological Profile .................................................. 67
Mode of Action and Direct Insecticidal Activity .................... 68

Indirect Insecticidal Effects ...................................... 76

Resistance of Sucking Pests Against Imidacloprid ................... 78

Toxicology and Pharmacology of Imidacloprid ........................ 81

Symptomatology of Insects .......................................... 81
Pharmacology of Imidacloprid ....................................... 82
Pharmacokinetics and Metabolism of Imidacloprid .................... 83
Toxicology of Imidacloprid .......................................... 85
Toxicology of Imidacloprid 10% Spot-on ............................. 87
Tolerability of Imidacloprid 10% Spot-on in Dogs and Cats .......... 88
Ecological Effects and Ecotoxicological Studies .................... 92
Safety Assessment for Humans ....................................... 94

Imidacloprid as a Veterinary Product for Flea Control .............. 98

Clinical Efficacy of Imidacloprid (Advantage®) in Dogs and Cats........... 106
  *Laboratory Studies in Dogs* ......................................... 108
  *Field Studies in Dogs* ............................................... 113
  *Laboratory Studies in Cats* .......................................... 120
  *Field Studies in Cats* ................................................ 125
Comparative Studies (Between Imidacloprid
and Other Compounds for Flea Control) ............................. 127
  *Laboratory Studies* .................................................. 127
  *Field Studies* ........................................................ 136
Larvicidal Effect of Imidacloprid
and Efficacy Enhancement (with PBO) ............................... 141
  *Larvicidal Effect* ................................................... 141
  *Efficacy Enhancement* ............................................... 148
Environmental and Habitual Factors
Influencing Imidacloprid Treatment ................................... 149
  *Effects of Shampooing and Repeated Water Exposure* .................. 149
'Umbrella Effect' ....................................................... 153
Small Domestic Animals ................................................ 154
  *Rabbits* ............................................................. 154
  *Ferrets* ............................................................. 157
  *Rodents and Other Small Animals* .................................... 159
Ectoparasitic Spectrum of Activity ..................................... 160
  *Louse Infestation* ................................................... 160
  *Sheep Ked Infestation* ............................................... 161

**References** ......................................................... 163

**Subject Index** ...................................................... 185

# Abbreviations

| | |
|---|---|
| **ACh** | acetylcholine |
| **AChR** | acetylcholine receptor |
| **a.i.** | active ingredient |
| **α-BGT** | α-bungarotoxin |
| **b.w.** | body weight |
| **CBH** | cutaneous basophilic hypersensitivity |
| **cDNA** | complementary desoxyribonucleic acid |
| **CNS** | central nervous system |
| **dpt** | days post treatment |
| **e.g.** | lat. exempli gratia – for instance |
| **EM** | electron microscopy |
| **FAD** | flea allergic/allergy dermatitis |
| **GABA** | γ-aminobutyric acid |
| **HCH** | hexachlorocyclohexane |
| **IDI** | insect development inhibitor |
| **i.e.** | lat. id est – that is to say |
| **IGR** | insect growth regulator |
| **IDST** | intradermal skin test |
| **IPM** | integrated pest management |
| **IRAC** | insecticide resistance action committee |
| **IRM** | integrated resistance management |
| **LC$_{50}$** | lethal concentration with 50% death |
| **LD$_{50}$** | lethal dosage with 50% death |
| **mAChR** | muscarinic acetylcholine receptor |
| **mPa** | milli-pascale |
| **nAChR** | nicotinic acetylcholine receptor |
| **NOEL/NOEC** | no observed effect level/concentration |
| **NMH** | nitromethylene heterocycle |
| **MD** | miliary dermatitis |
| **MFO** | mixed function oxidase |
| **MOS** | margin of safety |
| **p.a.** | post application |
| **PBO** | piperonyl butoxide |
| **p.t.** | post treatment |

| | |
|---|---|
| **ppm** | parts per million |
| ® | registered trademark |
| **RfD** | reference dose |
| **RH** | relative humidity |
| **SAR** | structure-activity relationship |
| **SD** | standard deviation |
| **TEM** | transmission electron microscopy |
| ™ | trademark |
| **WHO** | World Health Organization |
| **wt/wt** | weight per weight |
| **w/v** | weight per volume |

# Acknowledgements

The authors wish to acknowledge the support of the following persons
(in alphabetical order):

*Dr. Peter Andrews, David Bledsoe DVM, Ulrike Diez, Dr. Christian Epe, Waltraud Erasmy,
Dr. Vincent Feyen, Ronan Fitzgerald DVM, Stefan Freeman, Dr. Laurie Griffin,
Dr. Olaf Hansen, Dr. Terry Hopkins, Dr. Peter Jeschke, Jürgen Junkersdorf,
Dr. Ralph Krebber, Dr. Brigit Laber-Probst, Dr. Wolfgang Leicht, Prof. Heinz Mehlhorn,
Martina Mencke, Dr. Ralf Nauen, Dr. Reiner Pospischil, Barbara Prosten,
Dr. Michael Rust, Dr. Klaus Reuter, Prof. Thomas Schnieder, Iris Schröder,
Dr. Carsten Staszyk, Dr. Andreas Turberg, Dr. Giampiero Vantellino, Roswitha Wolff*

# Introduction

Fleas are wingless insects with a laterally compressed body of about 1.5-4 mm length. Like all insects they possess six legs and three body segments. Taxonomically they belong to the order Siphonaptera (Eckert et al. 2000) (Table 1). This family contains several species and subspecies. Fleas represent one of the most important ectoparasites (Mehlhorn 2000; Mehlhorn et al. 2001b). At the moment there are more than 2000 described species and subspecies throughout the world (Borror et al. 1981). These species belong to the families Pulicidae, including *Pulex* spp., *Ctenocephalides* spp., *Spilopsyllus* spp. and *Archaeopsyllus* spp., or the familia Ceratophyllidae with the genuses *Ceratophyllus* or *Nosopsyllus* to mention only some of the most important veterinary and human representatives.

Fleas have a history of about 60 million years and were already found on prehistoric mammals. While becoming parasitic the original exterior of the two-wing insects, also designated as the order Diptera, has changed by losing the wings in the adults, whereas the larval form still has similarity with the larva of the order Diptera (Strenger 1973).

About 95% of the ~2000 different flea species parasitize on mammals, 5% live on birds.

Table 1. Taxonomy of fleas

| Systematic Taxonomy | | | |
|---|---|---|---|
| Phylum | | | Arthropoda |
| Subphylum | | | Tracheata (=Antennata) |
| Classis | | | Insecta (Hexapoda) |
| Ordo | | | Siphonapterida |
| Familia | Pulicidae | Familia | Ceratophyllidae |
| Genus | *Ctenocephalides, Pulex, Spilopsyllus, Archaeopsyllus* | Genus | *Ceratophyllus, Nosopsyllus* |
| Species | *C. felis* (Bouché 1835) *C. canis* (Curtis 1826) *P. irritans* (Linné 1758) *S. cuniculi* (Dale 1878) *A. erinacei* (Bouché 1835) | Species | *C. gallinae* (Schrank 1804) *C. columbae* (Stephens 1829) |

# General Morphology

The flea is dark brown in color, wingless and possesses a laterally compressed chitineous abdomen (Soulsby 1982). The glossy surface of the body allows easy movement through hair and feathers (Urquhart et al. 1987). Compound eyes are absent, but some species have large or small simple eyes. The legs are long, strong and adapted for leaping (Soulsby 1982). This can especially be seen in the third pair of legs which is much longer than the others (Urquhart et al. 1987) and muscular. In some species there are a number of large spines on the head and the thorax known as 'combs' or ctenidia. There may be a genal comb on the cheek (gena) and a pronotal comb on the posterior border of the first thoracic segment (Soulsby 1982). These combs or ctenidia belong to one of the three sets of characteristics in morphological taxonomy for identifying fleas, the so-called chateotaxy. Thoracic and leg structures and the structure of the male segment IX, the female sternite VII and the sperm holding organ (spermatheca) are the other two characterizing sets (Ménier and Beaucournu 1998).

Concentrating on pets, particularly on cats and dogs, only a restricted number of flea species occur in large amounts with any regularity to be of importance as nuisance pests. For the USA these are *Ctenocephalides felis felis*, the cat flea, *Ctenocephalides canis*, the dog flea, *Pulex irritans*, the human flea, and *Echidnophaga gallinacea* as well as *Ceratophyllus gallinae*, fleas found on poultry (Kalkofen and Greenburg 1974; Amin 1976; Harman et al. 1987; Dryden 1988). Similar situations are also found in Europe and other parts of the world with *C. felis felis* and *C. canis* mainly dominating and *P. irritans* and *Archaeopsylla erinacei*, the flea of the hedgehog, as species of possible high rates of infestations (24% (Baker and Hatch 1972) and 8.3% (Kristensen et al. 1978)) (Fig. 1).

As can be seen in the chapter 'Origin and Distribution' the species *Ctenocephalides* with its representatives *C. felis felis* and *C. canis* both have a worldwide distribution (Kalvelage and Münster 1991) and are the most important flea species parasitizing dogs and cats as man's most wide-spread companions. Because of their genus similarities as well as their similar pattern of distribution and host spectrum, it is presumed that their biology does not show essential differences (Kalvelage and Münster 1991). Due to the great variety in species, hosts, environmental pattern and biology we will focus on the cat and dog flea particularly, the cat flea *C. felis felis* (Bouché 1835) being the most important ectoparasite of companion animals such as cats and dogs (Rust and Dryden 1997).

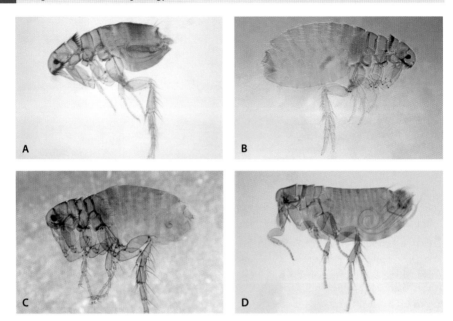

Fig. 1. A-D. Morphology of **A** Cat flea *Ctenocephalides felis felis* (Bouché 1835); **B** Dog flea *Ctenocephalides canis* (Curtis 1826); **C** Hedgehog flea *Archaeopsylla erinacei*; **D** Poultry flea *Ceratophyllus gallinae* (Schrank 1802)
original size **A:** 2.1 mm; **B:** 3.2 mm; **C:** 3.2 mm; **D:** 2.2 mm

# Taxonomy

**3**

In *C. felis* there are four recognized subspecies throughout the world, all of which are primarily parasites of carnivores (Lewis 1972; Hopkins and Rothschild 1953). *C. felis damarensis* and *C. felis strongylus* are restricted to Africa, *C. felis orientis* is found in southeast Asia and the East Indies (Lewis 1972), primarily infesting cattle, sheep and goats whereas the first two are found as parasites of wild carnivores (Dryden 1993). *C. felis felis* is found worldwide on many species of wild and domesticated animals (Rust and Dryden 1997). It is the only subspecies that occurs in North America (Dryden 1993) and is often only referred to as *C. felis,* also throughout this book. *C. felis felis* was probably introduced quite recently into Europe when domestic cats were imported at the time of the Crusades (Petter 1973; Beaucournu 1990). Believed to originate from Africa, the so-called 'cat flea' *C. felis* is now cosmopolitan, ranging from warm tropical areas to temperate zones with prolonged subfreezing temperatures (Lewis 1972).

New investigations in the field of taxonomic differentiation by using the phallosome structures as identification key propose the status of *C. orientis* and *C. damarensis* as a full species (Ménier and Beaucournu 1998) and confirm among others the findings of Haeselbarth (1966) concerning *C. orientis* and Beaucournu (1975) concerning *C. damarensis* against all controversies. The morphological differentiation between *C. felis* and *C. canis* as well as some other major flea species is given in Fig. 2, Fig. 3 and Table 2.

# Pictural key to the genus *Ctenocephalides*

Genus *Ctenocephalides* – general description

Genal and pronotal combs present

Genal comb = 7 – 9 pointed spines
Pronotal comb = 7 – 9 spines on each side
Body lenght = 1.5 – 3.2 mm
Colour – brown
Eyes – present

## Pictural key to Dog and Cat fleas

| | *Ctenocephalides felis* | | *Ctenocephalides canis* | |
|---|---|---|---|---|
| | Male | Female | Male | Female |
| Shape of head capsule | | | | |
| Spine 1 and 2 of the genal comb | Both 1st and 2nd spine have the same lenght | | 1st spine is half as long as 2nd spine | |
| Number of teeth of tibiae | Tibiae of all 6 legs have 4 to 5 teeth | | Tibiae of all 6 legs have 7 to 8 teeth | |

**Fig. 2.** Morphological differentiation of the cat flea (*C. felis*) and the dog flea (*C. canis*) original size upper right: 3.0 mm

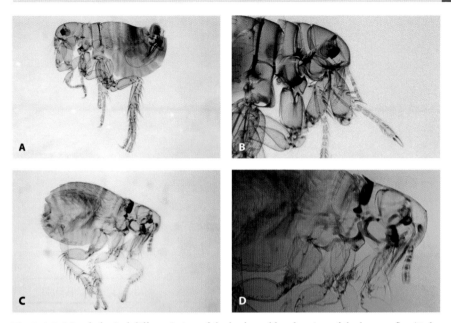

Fig. 3. A-D. Morphological differentiation of the body and head region of the human flea (*Pulex irritans*) (**A, B**) and the rabbit flea *(Spilopsyllus cuniculi)* (**C, D**) (details see Table 2) original size **A**: 2.5 mm; **C**: 1.8 mm

Table 2. General morphological differentiation using the presence or absence of pronotal and genal combs in fleas

| Without Combs present | Only pronotal combs present | Pronotal and genal combs present | |
|---|---|---|---|
| | | Only few combs present | Several Combs |
| *Xenopsylla cheopis* Rothschild 1903 (Oriental rat flea) | *Nosopsyllus fasciatus* Bosc 1800 (Northern rat flea) | *Ischnopsyllidae* (Bat fleas) | *Spilopsyllus cuniculi* Dale 1878 (European rabbit flea) |
| *Pulex irritans* Linnaeus 1758 (Human flea) | *Ceratophyllus gallinae* Schrank 1802 (Poultry flea) | *Archaeopsylla erinacei* Bouché 1835 (Hedgehog flea) | *Leptopsylla segnis* Schönherr 1816 (European mouse flea) |
| *Echidnophaga gallinacea* (Sticktight flea) | *Diamanus montanus* (Ground squirrel flea) | | *Ctenocephalides canis* Curtis 1826 (Dog flea) |
| | *Orchopeas howardii* (Squirrel flea) | | *Ctenocephalides felis* Bouché 1835 (Cat flea) |
| | | | *Cediopsylla simplex* (Common eastern rabbit flea) |

# Dissemination and Economic, Veterinary and Medical Importance

## Origin and Distribution

*C. felis* and *C. canis* are the two most important flea species of companion animals worldwide. Their pattern of distribution as found by different working groups is gathered into the following tables (Table 3A, Table 3B).

Table 3A. Distribution pattern of the cat flea (*C. felis*) taking into consideration the host animal, the geographical appearance and the rate of infestation (where mentioned)

| Location | Host Animal | Infestation Rate | Reference |
|---|---|---|---|
| Argentina | dog | most prevalent | Lombarddero and Santa-Cruz 1986 |
| Australia, Queensland | dog | most prevalent | Cornack and O'Rourke 1991 |
| Austria | dog | 81.4% prevalence | Supperer and Hinaidy 1986 |
| Austria mixed infestation | cat | 96.3% prevalence | Supperer and Hinaidy 1986 |
| Czechoslovakia | dog | 0.0% prevalence | Zajicek 1987 |
| Denmark | dog | most prevalent | Kristensen and Kieffer 1978; Kristensen et al. 1978 |
| Denmark | dog | 54.3% prevalence | Haarløv and Kristensen 1977 |
| Denmark | dog | 64.7% prevalence | Kristensen et al. 1978 |
| Denmark | cat | 100% prevalence | Haarløv and Kristensen 1977 |
| Denmark mixed infestation | cat | 95.3% prevalence | Kristensen et al. 1978 |
| Egypt | dog | most prevalent | Amin 1966 |
| France | dog | most prevalent | Bourdeau and Blumenstein 1995 |
| Germany | dog | most prevalent | Liebisch et al. 1985; Müller and Kutschmann 1985 |

| Location | Host Animal | Infestation Rate | Reference |
|---|---|---|---|
| Germany (West) | dog | 59.9% prevalence | Liebisch et al. 1985 |
| Germany (East) | dog | 45.8% prevalence | Müller and Kutschmann 1985 |
| Germany | dog | 1.2% prevalence | Kalvelage and Münster 1991 |
| Germany | cat | 11.1% prevalence | Kalvelage and Münster 1991 |
| Germany (West) mixed infestation | cat | 100% prevalence | Liebisch et al. 1985 |
| Great Britain | dog | 17.1% prevalence | Beresford-Jones 1981 |
| Great Britain | cat | 100% prevalence | Beresford-Jones 1974 |
| Great Britain | cat | 100% prevalence | Beresford-Jones 1981 |
| Ireland | dog | 4% prevalence | Baker and Hatch 1972; Baker and Mulcahy 1986 |
| Puerto Rico | dog | most prevalent | Fox 1952 |
| Poland | dog (strays) | only a few specimen found on dogs | Piotrowski and Polomska 1975 |
| South Africa, Western Cape | dog | most common | Briggs 1986 |
| South Africa, Transvaal | dog | most common | Horak 1982 |
| Sweden | dog | 0.0% prevalence | Persson 1973 |
| United Kingdom | dog | most prevalent | Beresford-Jones 1981; Chesney 1995; Coward 1991; Geary 1977 |
| United Kingdom, London | dog | most prevalent | Beresford-Jones 1981 |
| United Kingdom | dog | 15.4% prevalence | Beresford-Jones 1974 |
| USA, southeastern Wisconsin | dog and cat | most prevalent | Amin 1976 |
| USA, Indiana | dog and cat | most prevalent | Dryden 1988 |
| USA, Virginia | dog and cat | 100% prevalence | Painter and Echerlin 1985 |
| USA, North-Central Florida | dog | 92.4% prevalence | Harman et al. 1987 |
| USA, North-Central Florida | cat | 99.8% prevalence | Harman et al. 1987 |

Table 3B. Distribution pattern of the dog flea (*C. canis*) taking into consideration the host animal, the geographical appearance and the rate of infestation (where mentioned)

| Location | Host Animal | Infestation Rate | Reference |
|---|---|---|---|
| Austria | dog | most prevalent | Ressl 1963 |
| Austria | dog | 18.6% prevalence | Supperer and Hinaidy 1986 |
| Austria mixed infestations | cat | 3.7% prevalence | Supperer and Hinaidy 1986 |
| Czechoslovakia | dog | 2.0% prevalence | Zajicek 1987 |
| Denmark | dog | 42.1% prevalence | Haarløv and Kristensen 1977 |
| Denmark | dog | 29.8% prevalence | Kristensen et al. 1978 |
| Denmark mixed infestations | cat | 5.3% prevalence | Kristensen et al. 1978 |
| Germany (West) | dog | 42.9% prevalence | Liebisch et al. 1985 |
| Germany (East) | dog | 39.6% prevalence | Müller and Kutschmann 1985 |
| Germany | dog | 0.6% prevalence | Kalvelage and Münster 1991 |
| Germany (West) mixed infestations | cat | 5.3% prevalence | Liebisch et al. 1985 |
| Great Britain | dog | 6.2% prevalence | Beresford-Jones 1974 |
| Great Britain | dog | 2.6% prevalence | Beresford-Jones 1981 |
| Great Britain (rural) | dog | most prevalent | Edwards 1968 |
| Ireland | dog | most common | Baker and Hatch 1972; Baker and Mulcahy 1986 |
| Ireland, Dublin | dog | most prevalent | Baker and Hatch 1972 |
| Ireland | dog | 86% prevalence | Baker and Hatch 1972; Baker and Mulcahy 1986 |
| New Zealand | dog | most prevalent | Guzman 1984 |
| New Zealand | dog | most prevalent | Guzman 1984 |
| Poland | dog (strays) | 40% prevalence | Piotrowski and Polomska 1975 |
| Sweden | dog | 0.3% prevalence | Persson 1973 |
| USA, Indiana | dog | 18% prevalence | Dryden 1988 |

## Economic Importance

Flea infestation is the most common ectoparasitic condition of dogs and cats in North America (Rust and Dryden 1997). Fleas are also the most common parasites found on cats and dogs along the coast of Southern Africa (Briggs 1986), are a common ectoparasite of dogs and cats in Ireland (Baker and Elharam 1992) and possess a cosmopolitan distribution (see Tables 3A and 3B). This wide distribution and the fact that fleas are major nuisance pests (Dryden and Rust 1994), a matter of public health and the source of flea allergic dermatitis (FAD), one of the most common causes for the presentation of dogs to the veterinarian, make control definitely necessary. But control of the flea on pets and in the environment can be expensive, time consuming and often frustrating (Dryden and Smith 1994). The annual expenditures by pet owners for flea control products in the United States exceed $ 1 billion (Conniff 1995). And apart from expenditures on control measures, flea-related diseases account for over 50% of the dermatological cases reported to veterinarians and 35% of their total effort (Bevier-Tournay 1989; Kwochka 1987). This data emphasizes the necessity of an effective flea control if only from the economical point of view.

Market surveys in Animal Health companies showed that the ectoparasiticide market in companion animals was estimated to be around 1,1 billion Euro. This is about 30% of the total Companion Animal Market, while the total Animal Health Market is around 3,7 billion Euro. The market is dominated by the two major continents North America with a market share of around 70% and of 20% in Western Europe. The distribution of the ectoparasiticide market into the animal species dogs, cats and horses is given in Fig. 4.

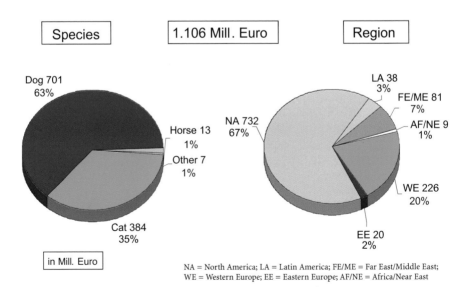

NA = North America; LA = Latin America; FE/ME = Far East/Middle East;
WE = Western Europe; EE = Eastern Europe; AF/NE = Africa/Near East

**Fig. 4.** Ectoparasiticide market in companion animals in 1999 by species and region

# Veterinary Importance

As already mentioned above ('Economic Importance') flea allergy dermatitis (FAD) is one of the most frequent causes of skin conditions in pets (Southern United States (Halliwell 1985a), Dublin (Baker and Hatch 1972)) and a major clinical entity in dogs, particularly in the warm and humid regions of the central states of North America (Gross and Halliwell 1985). Harman et al. (1987) even state FAD as the commonest non-routine reason for dog owners to seek veterinary advice in the USA (Harman et al. 1987).

The hypersensitivity to flea bites is not only seen in dogs but also a major cause of feline miliary dermatitis (Baker and O'Flanagan 1975; Halliwell 1979; Muller et al. 1983) and other pruritic skin diseases in cats (Scott 1980).

Besides the dermal aspect, fleas as hematophagous insects can also cause an unbalance in the circulatory system. Iron deficiency anemia in heavy infestations can be particularly produced in young animals. The genus *Ctenocephalides* has been reported to produce anemia in dogs, cats, goats, cattle, and sheep (Obasaju and Otesile 1980; Harvey et al. 1982; Soulsby 1982; Fagbemi 1982; Blackmon and Nolan 1984; Yeruham et al. 1989), which was even combined with mortality in calves, lambs and kids (Yeruham et al. 1989).

Female fleas are reported to consume an average of 13.6 µl (+/-2.7 µl) of blood per day (Dryden and Gaafar 1991). By that volume of blood, 72 female fleas could consume one milliliter of blood daily (Dryden 1993). According to Dryden and Gaafar (1991) an infestation of 220 female fleas could potentially consume 10.0% (i.e. 3.0 ml) of an 0.45 kg kitten's blood per day, which would cause anemia in the kitten. Thus, a heavy flea infestation may rapidly produce anemia in a small puppy or kitten (Dryden and Gaafar 1991). But in mature cats, dogs or young livestock with a comparatively greater blood volume and iron store it is more likely that very high infestation levels must be maintained over several weeks to produce anemia from chronic blood loss and that the reported cases of extreme blood loss (apart from very young animals) are likely to be combined with other factors such as a systemic reaction to antigenic components in the flea saliva, stress or unknown factors e.g. impairing iron utilization (Dryden and Gaafar 1991). Whether alone or in conjunction with other factors, the blood consumption caused by heavy flea infestations is nevertheless immense.

Furthermore, the presence of fleas on pets indicates an insufficient ectoparasite control regime in that the pet is also not protected against and susceptible to cheyletiellosis, pediculosis, otodectic otitis, scabies, and even babesiosis from tick infestation (Briggs 1986).

The last aspect of veterinary importance of flea infestations is the role as transmitter of a variety of diseases which also represent a potential health risk for humans (see chapter 'Medical Importance').

Important for the host is the infection with the cestode *Dipylidium caninum* by ingesting infested fleas and the infection with the subcutaneous filarid nematode *Dipetalonema reconditum* as a non-pathogenic filarid which must be considered as differential diagnosis of the microfilariae of the pathogenic dog heartworm, *Dirofilaria immitis* (Georgi 1990).

In a study in Vienna by Hinaidy (1991) every 44th flea from cats and 61st flea from dogs harbored cysticercoids of *D. caninum*. The infection intensity rates ranged between 2.3% for *C. felis* in cats, 1.2% for *C. felis* in dogs and 3.1% for *C. canis* in dogs with male fleas being more extensively, but less intensively infected than female fleas.

## Medical Importance

In general there have been risks and health problems connected with arthropods (Genchi 1992). Concerning fleas, their readiness to parasite humans as alternative hosts gives the fleas of domestic animals a relevance in public health (Urquhart et al. 1987). The polyxenous state of *C. felis* and *C. canis* has been recorded in several surveys (Baker and Hatch 1972; Kristensen et al. 1978; Beresford-Jones 1981; Kalvelage and Münster 1991; Chesney 1995) and verifies the possibility of a transfer of various agents between animals particularly pets and humans. *C. felis* has quite a low host specificity and is able to feed on humans (Genchi 1992). In towns it maintains its life cycle (mainly) indoors feeding on pets. Populations grow throughout the year in buffered microclimate such as buildings in towns, and show unexpected large peaks caused by the sudden and synchronous breakdown of nymphal diapause of a very large number of individuals when the habitations are abandoned by domestic animals and their owners for longer or shorter times. Human infection by *C. felis* is not only occasional, but quite frequent at that stage (Genchi 1992). But in general fleas only use humans as an alternate host when the flea population builds up to a critical level or when an infested animal leaves the shared habitation for an extended period (Pullen and Meola 1995). Intense pruritus can then occur as one of the effects seen in children sensitized to flea bites (Service 1980).

C. felis and C. canis play an important role as transmitters of a wide spectrum of diseases and have been reported to be the intermediate host of *D. caninum*, the dog tapeworm (Soulsby 1982; Chen 1933, 1934; Joseph 1974; Joyeux 1916; Marshall 1967; Pugh 1987; Pugh and Moorhouse 1985; Venard 1938; Yasuda et al. 1968; Yutuc 1968), *Hymenolepis nana* (Bacigalupo 1931; Marshall 1967), *H. diminuta* (Joyeux 1916; Marshall 1967), *H. citelli* (Marshall 1967), *H. microstoma* (Marshall 1967) and *Dipetalonema reconditum* (Farnell and Faulkner 1978; Hinaidy et al. 1987; Nelson 1962; Nelson et al. 1962; Pennington and Phelps 1969) as further helminths. Beside these agents they are also reported to be the transmitter of Friend Leucemia Virus (Rehacek et al. 1973), *Rickettsia typhi* (Farhang-Azad et al. 1984; Irons et al. 1946; Keaton et al. 1953), *Rickettsia* sp. (Stephen and Rao 1980), *Yersinia pestis* (Blanc and Baltazard 1914; Ioff and Pokrovskaya 1929; Wheeler and Douglas 1941), *Pasteurella* sp. (Sapre 1946), *Brucella melitensis*, *Br. abortus*, *Br. suis* (Tovar 1947) as well as the mites *Cheyletiella parasitivorax* (Rak 1972) and *Cheyletiella* sp. (Guzman 1982). Of those agents most important for man are *D. caninum*, which is an occasional parasite of man (Service 1980), and, of greater concern, the murine typhus (*Rickettsia thyphi*) (Traub et al. 1978) and plague (*Yersinia pestis*) (Pollitzer 1960).

Murine typhus is a mild febrile disease characterized by development of headaches, chills, and skin rashes, with infrequent involvement of the kidneys and the central nervous system (Azad 1990). It occurs in humans and many small animals, including rats and mice (Dryden and Rust 1994). Its primary vector was traditionally believed to be the oriental rat flea, *Xenopsylla cheopis* (Dryden and Rust 1994). The disease occurs in the United States along the southeastern, southwestern and gulf coasts. In southern California, it appears that the principal transmission cycle involves opossums and *C. felis* (Adams et al. 1970).

An additional typhus-like rickettsiae (ELB agent) was identified by Williams et al. (1992), harboring in opossums and *C. felis* in southern California whose role in the epidemiology of murine typhus is so far unknown.

Concerning cat fleas as a vector for the etiologic agent of bubonic plague, the dog has been shown to be capable of carrying *Yersinia pestis* (Rust et al. 1971). Human plague cases and even deaths associated with infected cats have been occasionally reported (Kaufmann et al. 1981; Werner et al. 1984; Eidson et al. 1988). The chances of contact with plague-infected fleas are increasing with humans and pets encroaching upon endemic wildlife plague cycles. Humans can become infected by their pets via pneumonic transfer, bites from infected rodent fleas residing temporarily upon the pets or possibly via cat flea transmission (Dryden and Rust 1994).

# Developmental Cycle of Fleas

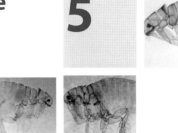

The flea develops via a number of stages, beginning with the egg, followed by the larva, pupa and finally adult stage. The life cycle of the flea is one of complete metamorphosis. It can be completed in as little as 14 days or be prolonged up to 140 days, depending mainly on temperature and humidity (Silverman et al. 1981b). The life cycle of most flea species is characterized by three events: the hatching of the egg, the period from 1st instar to pupa, and the period from pupa to adult (Linardi et al. 1997) (Fig. 5).

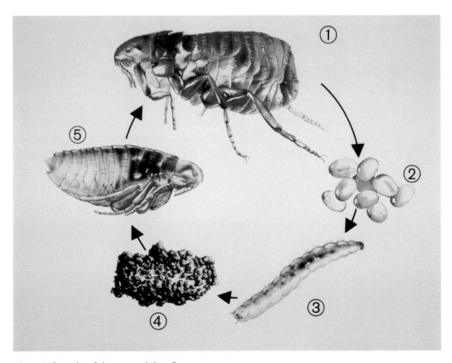

**Fig. 5.** Life cycle of the cat and dog flea
①Adult female flea ② Eggs ③ Larva ④ Pupa ⑤ Preemerged adult

# Eggs

Cat flea eggs possess a widely oval form, rounded at both ends, a slightly transparent color at the beginning, later a pearly white color, 0.5 x 0.3 mm size, a smooth surface which can slightly darken later on (Karandikar and Munshi 1950), and they are well visible to the naked eye (Fig. 6). They are laid nearly exclusively on the host by mature females and do not immediately fall from the animal after they are positioned, according to Rust and Dryden (1997). Initially the chorion of the egg is wet, which tends to prevent drop-off. But it dries rapidly and 60% of the eggs drop off within two hours of deposition (Rust 1992). The rate at which eggs drop or are dislodged from the pelage is influenced by grooming, hair coat length, and host activity (Rust and Dryden 1997). Within eight hours about 70% of them are dislodged from the host (Rust 1992).

Temporal patterns of egg deposition are reported with a correlation between peak egg production and periods of the cat sleeping or resting at the middle of scotophase. Significantly fewer eggs are produced near dusk and dawn (Kern et al. 1992). Rust (1992) reported similar results of peak egg production during the last eight hours of scotophase, but no correlation between peak egg production and resting or sleeping patterns in test cats. Furthermore fleas did not change their egg production in response to different photoperiods (Metzger and Rust 1996). Dropping off the host, flea eggs accumulate in areas where pets sleep and rest (Byron 1987).

Egg deposition by adult females does not take place during the first 24 hours of blood feeding and is only less than half of the daily average during the second 24 hour period (Thomas et al. 1996). Further a low average fecundity was recorded during the first three days (Thomas et al. 1996). The start of the egg production is also influenced by the age of the female flea: Osbrink and Rust (1984) reported of 4-day-old adult females placed on a host, beginning egg production after two days, whereas adult females younger than 24 hours require three to four days on the host before egg production starts (Osbrink and Rust 1985b). On cats, fleas reach reproductive maturity within the first week, with a peak of egg output occurring sometimes as quickly as within three days (Williams 1983). But it may also take up to nine days for them to reach maximum egg output (Williams 1983; Osbrink and Rust 1984; Dryden 1989b). In general, it is suggested that fleas do not mate before bloodfeeding or if they do, the

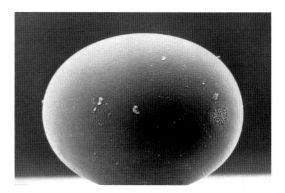

**Fig. 6.** EM picture of a flea egg (original size 0.5 x 0.3 mm)

mating is unsuccessful. In the cat flea, bloodfeeding is apparently necessary for both oviposition and successful mating. Mating, however is not necessary for oviposition (Zakson-Aiken et al. 1996).

The number of eggs collected in different studies to gain an impression of daily production and total egg count varies greatly between hosts (Dryden 1989b; unpublished data in Rust 1994). Rust (1994) presumes possible reasons for this to be differences in grooming which greatly affects adult survival (Dryden 1988, 1989b; Wade and Georgi 1988) or possible differences in the host's physiology. The reproduction in fleas is also stimulated by estrogens and corticosteroids in the peripheral circulation of the host as shown by Rothschild and Ford (1964) for the rabbit flea, possibly explaining individual variation in the extent to which a host attracts fleas (MacDonald 1984).

Adult female fleas may produce from eleven to 46 eggs per day (Osbrink and Rust 1984; Dryden 1988, 1989b; Hink et al. 1991). According to Hinkle (1992) fleas produce an average of 24-25 eggs per female per day on cats. Osbrink and Rust (1984) suggested that females might deposit between 300 and 500 eggs in their lifetime with an average of 158.4 eggs per female flea. Patton (1931) and Smit (1973) reported of 800 to 1000 eggs per female flea. Dryden (1989b) proposes a possible production of up to 1,745 eggs during a 50 day period and well over 2000 eggs over 113 days in unconfined fleas and cats restricted from grooming. According to him the cat flea is capable of producing 40-50 eggs per day during peak egg production, averaging 27 eggs per day over 50 days, with continued egg production for over 100 days. About 1.05-times the body weight is produced in form of eggs within 24 hours (Dryden and Gaafar 1991).

Thus, fleas are highly reproductive and work with a calculated loss as several other parasites do (Strenger 1973). Possibly because of the inability of the immature stages, particularly the larvae, to withstand extreme changes in temperature and humidity or because of a failure of eggs to be deposited in favorable microhabitats, the cat flea has adapted by having a large reproductive capacity (Dryden and Rust 1994).

Flea egg production is reported to be influenced by the source of host: fleas raised on cats produced five to nine times more eggs than those raised in an artificial system with a significant difference in the total number of eggs and the number of eggs per female (Wade and Georgi 1988). Furthermore cat fleas have been reported to feed and survive for some time on human hosts, but no viable egg production could be recorded (Bacot 1914; Joseph 1976; Akin 1984; Williams 1986), instead of viable eggs on a variety of other hosts - calves, hamsters, rats and dogs (Williams 1986). In contrast, Tränkle (1989) reported of reduced numbers of viable eggs produced by the cat flea when fed on human hosts. *C. canis* already failed to complete its life cycle on cats as no eggs laid developed to adults (Baker and Elharam 1992).

As is the case for all the flea's life stages the hatching of the egg is strongly influenced by environmental conditions (temperature and relative humidity (RH)). At 16°C the number of flea eggs hatching increased with rising RH from about 70% hatch at 33% RH up to 100% hatch at 92% RH. At 27°C nearly all eggs hatched when there was 50% RH or more. However, at 35°C the hatch only took place in moist air (75-92% RH). At the same high temperature moisture conditions below 75% RH caused desiccation. The reason for the failure of eggs to hatch in warm saturated air (i.e. 35°C) may be due to an accumulation of heat within the egg (Silverman et al. 1981b). Olsen (1985) found about 70% hatch when eggs were held at 24+/-1°C and 65+/-5% RH.

The time required for hatching increases from 1.5 to 6.0 days as temperature decreases from 32 to 13°C (Silverman et al. 1981b). An exposure to 3°C for one day kills 65% of the eggs. Longer exposures provide complete kill (Silverman and Rust 1983).

Eggs from *C. canis* hatched at 22°C and 25°C and at 50% and 75% RH (Baker and Elharam 1992). As reported for *C. felis* temperature has a dominant effect on the time taken for hatching. As temperature decreased the hatching time was observed to increase in a non-linear fashion by Baker and Elharam (1992) as well for *C. canis*, with RH being of greater importance for the (whole) development of *C. canis* than temperature. Summary see Box 1.

---

**BOX 1. Cat flea eggs**

- **Oval shape with a white shiny ivory surface**

- **Size 5 mm in length**

- **Appear 24 to 36 hours after first blood meal**

- **Average 27 ova per day**

- **Favorable condition: relative humidity >50%, temperature around 25° C**

---

# Larvae

Most flea species parasitize nest-dwelling animals, and the great majority of flea larvae live in the nest or den of their hosts (Marshall 1981). Among the nest-inhabiting flea larvae, there is a gradation of dependence on the host and on adult fleas for nutrition (Moser et al. 1991). In cat fleas, both larvae and adults are dependent on the blood of the host, and the larvae can be determined as obligate parasites (Dryden 1989b).

Newly hatched flea larvae are slender, white, apod (i.e. without feet), sparsely covered with short hair, two to five millimeters in length, and posses a pair of anal struts (Dryden 1993). Their body consists furthermore of a yellow-to-brownish head (Dryden 1989a) three thoracal (breast) segments and ten abdominal (belly) segments (Kalvelage and Münster 1991). The larvae have chewing mouth parts (Urquhart et al. 1987) and are free-living (Dryden 1993) (Fig. 7, Fig. 8). As the larvae have no feet they move by using their skin muscle tube on dry surface, managing to move quite rapidly. They are only able to stop by using the mouth parts and less effectively the soft push of the last segment.

The larva of the cat and dog flea furthermore passes through two molts, thus having three larval instars, and the third larval instar pupates. All instars have a two-hooked caudal anal process and 13 body segments (Harwood and James 1979). The first larval

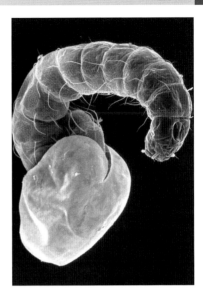

**Fig. 7.** EM picture of flea larva hatching from the egg (original size ~ 2 mm)

instar is approximately 2 mm in length, and the third instar can be 4 to 5 mm long (Elbel 1951; Bacot and Ridewood 1915).

The larval development occurs in protected microhabitats that combine moderate temperatures, high relative humidity and a source of nutrition in form of adult flea fecal blood (Dryden and Rust 1994). Cat flea larvae have a minimal nutritional requirement of dried blood to develop (Rust and Dryden 1997). They feed on adult flea feces which are under natural circumstances sufficient for living and development (Strenger 1973). The so-called debris usually thought to be an important part of larvae nutrition have no importance whatsoever for nourishment of the larvae (Strenger 1973). Künkel first reported in 1873 of this larval form of fecal nutrition, but believed it to be supplemented with other organic material digested by cat flea larvae (and dormouse flea larvae). Feeding trials by Strenger (1973) have proven that organic material is not ingested by the cat flea larvae. Apart from adult flea fecal blood, no organic material with the exception of flea eggs as well as injured flea larvae was proven to be ingested by cat flea larvae, this as an example of a form of cannibalism.

Once the larvae start feeding on adult feces the color of their gut which can be seen from outside turns into ruby-red (Strenger 1973), giving the larvae on the whole a brownish color.

The larvae are negatively phototactic (i.e. they move away from light), positively geotactic (i.e. they follow gravitation) (Byron 1987) and thigmotactic (i.e. they recognize tactile stimulus and react to mechanical contact) (Strenger 1973). This allows the larvae to find safe hiding places and protection against desiccation (Strenger 1973). Last but not least, larvae orient to sources of moisture, suggesting some type of hygrotactic response (Byron 1987). All these abilities allow the larvae to avoid direct sunlight in their microhabitat and make them actively move deep into carpet fibers or under organic

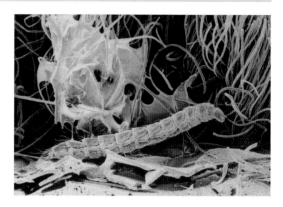

Fig. 8. EM picture of larva in its environment

debris (grass, branches, leaves, or soil) (Dryden 1993). In particular the debris which was formerly thought to be of great importance for larval nourishment plays an important role as substratum to satisfy the tigmotactic abilities of the flea larvae making them stay in a safe protected place (Strenger 1973).

About 83% of fleas develop at the base of carpets in the home (Byron 1987). Larvae are capable of moving at least 46 cm in carpet (Osbrink, personal communication, cited in Rust and Dryden 1997). Kern (1991) similarly reported that larvae would crawl several inches to avoid light and were observed to burrow an average of 2.36 mm into the sand, reaching a maximum depth of 7.5 mm. But despite all this ability to move, Byron's (1987) observation is still to be supported that first instars do not move far from the point of eclosion. The dispersion of the immature stages is mainly governed by the habits of the host (Dryden and Rust 1994).

As mentioned above larvae feed on adult flea feces. Adults guarantee for a sufficient nutrition of the larvae by imbibing much more blood from their host than they can use (Moser et al. 1991). Not only the amount of blood consumed, but also the obligate blood meal before every egg deposition means that nutrition for the offspring is guaranteed (Strenger 1973). Faasch (1935) reported that the imaginal flea gut can only contain 5 mm$^3$, but that several times as much is consumed and excreted as blood feces.

Fecal nutrients would be more efficiently utilized and nutrient investment selected for, if a female invested nutrients in her own progeny (i.e. more eggs or more nutrient filled eggs) rather than investing equally in the progeny of conspecifics. However if most larvae are related to the adults that are feeding them, there is a strong selective advantage to nutrient excretion, for it is likely, at least in fleas that inhabit the nests of their hosts, that all of the individuals are related closely (Silverman and Appel 1994).

Silverman and Appel (1994) propose an alternative hypothesis for blood excretion by *C. felis*: It could be a form of harvesting key nutrients generally found at low levels in blood by the adults, such as B vitamins, but then nevertheless concentrating a limited resource may be accomplished for a selfish reason, the offspring would still profit from it and the outcome may be vital for the offspring.

The incomplete utilization of host blood by adult fleas, the excretion of protein- and iron-rich feces, and the benefit derived from larvae consuming adult feces strongly indicates a unique form of parental investment in C. felis (Silverman and Appel 1994), with hemoglobin providing iron for normal growth and proper sclerotization as adults, and with serum including all the essential proteins (Moser et al. 1991). Fleas produce about 0.77 mg of feces per day without any particular temporal pattern (Kern et al. 1992). A spiral of feces is formed from the anus ten minutes after feeding (Akin 1984).

The larvae of the cat flea is capable of using blood from different hosts (Linardi et al. 1997; Linardi and Nagem 1972; Moser et al. 1991).

Silverman and Appel (1994) observed unfed larvae dying about three days after eclosion.

The larval stage of the cat flea usually lasts five to eleven days, depending on the climate and the availability of food (Lyons 1915; Silverman et al. 1981b).

In the development and survival of newly hatched larvae relative humidity is a major influencing factor (reported here for C. canis) (Baker and Elharam 1992). At 24.4°C and 78% RH, pupation of a larval cohort began on day 7 and was complete by day 11 (Dryden 1988). As the temperature decreases, the length of time for larval development increases (Dryden 1993).

Bruce (1948) found that cat flea larval survival was >90% at temperatures of 21-32°C, but survival dropped to 34% at 38°C. Furthermore he reported of no larval survival at optimal temperatures, but RH of <45% or >95%. Relative humidities of 65-85% resulted in >90% larval survival. And larvae raised at 50% RH had development times twice as long as larvae raised at 65-85% RH. Cat flea larvae as well as pupae did not survive temperatures of >35°C for >40 hours/month, when the relative humidity was held constant at 75% (Silverman and Rust 1983). They also reported that only 7% of larvae survived a 1-day exposure at -1°C (75% RH) and that 43% of third instars exposed to 12% RH survived for 24 hours at 16°C and 97% survived at 10°C. Larvae are more sensitive than eggs to low humidity according to Silverman et al. (1981b). In their experiments all larvae died at any temperature if humidity was 33% RH or lower. All of the larvae could tolerate temperatures up to 27°C if the RH was at least 50%. High temperatures (35°C) had similar effects on the larvae as on the eggs with larvae desiccating at low RH and apparently overheating in saturated air (Silverman et al. 1981b). An exposition of five respectively 20 days at 3°C respectively 8°C is lethal for cat flea larvae (Silverman and Rust 1983). Pospischil (1995) reported of a larval development of 45 days at 15°C and first fleas hatching at that temperature after 70 days in contrast to a whole generation cycle of about 18 days at the temperature optimum between 25 and 30°C. Below 15°C no larval development is reported to be possible (Pospischil 1995). Concerning RH the same author reported of only 30% of the larvae finishing their development at 70% RH in contrast to >60% reproduction rate at a RH of 80-90%. The minimal RH which allows larval development is stated to be 60% (Pospischil 1995).

Kern et al. (1999) presume that larger larvae are apparently more resistant to desiccation than smaller larvae because of their smaller surface to volume ratio. They further believe differences in behavior towards temperature to be caused by variation between flea strains.

Outdoor survival is strongly influenced by temperature and humidity. Very high larval mortality was reported in sun-exposed areas (100%) and inside structures that trapped heat, such as the doghouse (100%) (Kern et al. 1999). Low larval survival has been reported at high temperatures, 37°C (Bruce 1948) and 35°C (Silverman et al. 1981b; Silverman and Rust 1983). A larval mortality of 100% has also been reported by 100% RH (Silverman et al. 1981b) respectively 95-100% RH (Bruce 1948).

Not only the nutritional requirements of larvae greatly limit the sites which are suitable for development, but also the necessity for humidity, with suitable outdoor sites even rarer (Dryden and Rust 1994). Flea larvae are not likely to survive outdoors in shade-free areas. Outdoor development probably occurs only in areas with shaded, moist ground, where flea-infested pets spend a significant amount of time to allow adult flea feces to be deposited into the larval environment. Likewise, in the indoor environment, flea larvae probably only survive in the protected environment under a carpet canopy or in cracks between hardwood floors in humid climates (Dryden 1993).

These developmental conditions verify that areas in close proximity to where pets sleep or rest need to be treated, because of the limited movement of the larvae (Dryden and Rust 1994). But there are still two reasons why flea larvae escape most adulticide treatments indoors: (1) The treatment fails to reach them at the base of carpet fibers where they develop; (2) about 2.5-times more insecticide per gram body weight is required to kill larvae than adults (Dryden and Rust 1994) (for further information see chapter 'Treatment of the Premises').

Summary see Box 2.

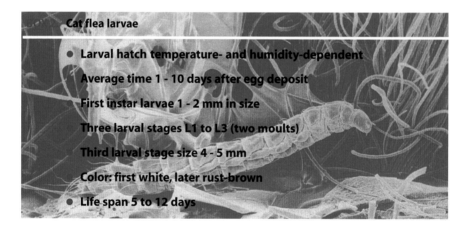

**Cat flea larvae**

- Larval hatch temperature- and humidity-dependent

  Average time 1 - 10 days after egg deposit

  First instar larvae 1 - 2 mm in size

  Three larval stages L1 to L3 (two moults)

  Third larval stage size 4 - 5 mm

  Color: first white, later rust-brown

- Life span 5 to 12 days

# Pupae

After completing development, the late third instar larva voids its alimentary canal contents in preparation of forming a cocoon and moves to an undisturbed place to spin a silk-like cocoon in which it pupates (Lyons 1915; Karandikar and Munshi 1950; Joseph 1981).

The principal factors triggering pupation are declining levels of juvenile insect hormone (Grant 1996).

The silkin excrete for the cocoon is produced by the salivary glands (Strenger 1973). The resulting cocoon consists of soft and moist silk-like material (Dryden 1993), measures about 4 by 2 mm (Soulsby 1982), is loosely spun and whitish in color (Dryden 1993) and is coated with dust and debris because of its stickiness (Soulsby 1982) which aids in camouflaging it perfectly (Karandikar and Munshi 1950; Dryden 1989a). Flea cocoons can be found in soil, on vegetation, in carpets, under furniture, and on animal bedding (Dryden 1993) (Fig. 9).

Within the pupal cocoon three distinct stages are found in the order Siphonaptera (to which the cat and the dog flea belong): The first is the U-shaped larval prepupa, the second is a true exarate pupa and the third is the preemergent adult, which has completed its pupal-imaginal molt but remains within the cocoon for varying lengths of time (Silverman et al. 1981b) (see chapter 'Preemerged Adults'). The U-shaped larva begins pupal development about 18 hours after the completion of the cocoon (Dryden and Smith 1994). If the larva is disturbed prior to this time in form of gentle shifting of the larval medium (e.g.) it might emerge and exit the cocoon. In trials by Dryden and Smith (1994) over 40% of those larvae did not spin a second cocoon, but developed as naked flea pupae. Metzger and Rust (1997) observed shorter times for adults to emerge from naked pupae compared with emergence time from cocoons.

The importance of this early emergence which can also be produced in carpeted homes by the vigorous agitation of the carpet by vacuums with rotary beater bars is so far unknown (Dryden and Smith 1994).

Of importance for a successful formation of the cocoon is the orientation against a perpendicular structure. Less than 3% manage to spin an enclosing cocoon in the

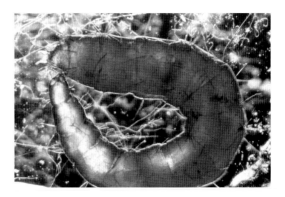

Fig. 9. EM picture of flea pupa

absence of a vertical surface, but over 95% of the larvae nevertheless survive to adulthood (Dryden and Smith 1994), thus demonstrating that the cocoon is not essential for the development of adults (Dryden and Smith 1994). Nevertheless it offers some benefits as e.g. protection from ant predation (Silverman and Appel 1984), somewhat impeding adult emergence and offering protection from non-host-produced stimuli and thus minimizing non-host-induced emergence (Silverman and Rust 1985). It does not provide a barrier to water loss (Silverman and Rust 1985) and also does not seem to be a barrier to insecticides (Dryden and Reid 1993), even though cocoons placed in a carpet appeared to be highly tolerant to a variety of insecticides (Rust and Reierson 1989). The survival of developing pupae in insecticide-treated households is caused by the lack of penetration of the carpet canopy by the insecticide (Dryden and Rust 1994).

Prepupae and pupae are fairly resistant life stages (Silverman et al. 1981b), but it is unknown how much of that protection is provided by the cocoon (Dryden and Smith 1994).

At 27+/-2°C, females pupate within about 32 hours and males pupate an average 12.1 hours later within about 44 hours (Dryden and Smith 1994). The degree of development within the cocoon at 16°C and 27°C was similar regardless of humidity (in trials of Silverman et al. 1981b). Under moderate conditions the pupa was the immature stage most resistant to desiccation; 80% survived to adulthood at 2% RH and 27°C (Silverman et al. 1981b). Exposure to 35°C during pupal development is uniformly lethal (Silverman et al. 1981b). Exposure to 3°C for five days and 8°C for 20 days is lethal as well (Silverman and Rust 1983).

Females develop into adults about 1.6 days faster than males (Dryden and Smith 1994). In trials of Metzger and Rust (1997) male prepupae and pupae developed more slowly than females at each temperature, requiring 14-20% more time to develop. At 15.5°C, the mean number of days for the female and male pupae to develop was 19.5 (female) and 23.5 (male), respectively (Metzger 1995), or even nine days earlier (Metzger and Rust 1997) and at 26.7°C, the mean number of days was 5.2 (female) and 6.5 (male), respectively (Metzger 1995), or two to five days earlier (Metzger and Rust 1997).

Not only females but also adults that developed from light prepupae in general remained in their cocoons for a shorter time than adults from more robust prepupae with two explanations proposed: either because of emergence mechanisms triggered when water and food reserves drop below a critical level or because of a weaker cocoon produced by incompletely nourished prepupa that does not impede adult emergence (Silverman and Rust 1985). When pupae were maintained at 24.4°C and 78% RH, adult *C. felis* began to emerge eight days after the initiation of pupal development and by day 13, all fleas emerged (Dryden 1988).

The pupa is suggested to be the stage most likely to survive extended periods in cool dry climates (Silverman et al. 1981b).

Summary see Box 3.

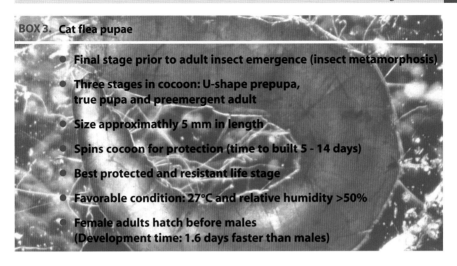

BOX 3. Cat flea pupae

- Final stage prior to adult insect emergence (insect metamorphosis)
- Three stages in cocoon: U-shape prepupa, true pupa and preemergent adult
- Size approximathly 5 mm in length
- Spins cocoon for protection (time to built 5 - 14 days)
- Best protected and resistant life stage
- Favorable condition: 27°C and relative humidity >50%
- Female adults hatch before males (Development time: 1.6 days faster than males)

## Preemerged Adults

Depending on the temperature inside the cocoon the flea develops via the stages prepupa and pupa within seven to 19 days into an adult which at first rests inside the cocoon (Karandikar and Munshi 1950; Silverman et al. 1981b). The observation of adult fleas remaining quiescent for prolonged periods within the pupal cocoon before emergence has been made by several researchers, including Bacot (1914), Karandikar and Munshi (1959) and Silverman et al. (1981b) and characterizes the so-called preemerged adult. The preemerged adult has a lower respiratory demand than the emerged adult and its survival is considerably longer under low humidity conditions (Silverman and Rust 1985). By that stage, prolonged adult survival within the cocoon, particularly under desiccating conditions, is possible and mainly due to quiescent periods of low metabolic activity rather than restriction of water loss through the cocoon wall (Silverman and Rust 1985). It can be suggested that the preemerged stage is ideal for prolonged survival during the absence of hosts or during unfavorable environmental conditions such as in winter or midsummer (Metzger and Rust 1997).

About 60% of adult fleas successfully emerge from cocoons held at 13°C by day 140 after eggs are collected (Silverman et al. 1981b). At 15.5°C, some adults emerge as late as 155 days (Metzger 1995). When held at 11°C and 75% RH adult *C. felis* may remain quiescent in the cocoon for up to 140 days (Silverman and Rust 1985). Exposure at 3°C for ten days and -1°C for five days was lethal to preemerged adults. Exposure at 8°C for 40 days resulted in 72% mortality. All preemerged adults held at 2% RH and 16°C could survive for >35 days, whereas 90% of the emerged adults would die (Silverman and Rust 1985). At 16°C in saturated air, 92% of the preemerged adults survived for 70 days compared with 62% of emerged adults (Silverman and Rust 1985). According to Rust and Dryden (1997) the decreased metabolic activity may in part explain the increased longevity of preemerged adults.

The option that the pupal stage can be as short as ten days, but the preemerged adults might remain in the cocoons for up to six months is meant by the term 'pupal window', which may cause problems in control measures and has to be understood by pet owners (Dryden 1996).

The ability to survive for extended periods in the cocoon is especially important for species such as *C. felis*, which infest mobile hosts that may not frequently return to a nest or burrow (Dryden and Rust 1994).

Fleas may continue to emerge from cocoons for up to four weeks after insecticide and insect growth regulator application to the environment (Dryden and Prestwood 1993). These resurgences are attributed to the presence of pupae and preemerged adults in cocoons at the time of treatment. Rust and Reierson (1989) noticed that preemerged adults in cocoons placed at the base of the carpet were not killed by sprays. Dryden and Rust (1994) believe the survival of pupae not to be due to any protection afforded by cocoons, but the lack of penetration of the carpet canopy by the insecticide, nevertheless causing the same problem of the pupal window in control measures.

Even though there is a direct relationship between temperature and rate of adult emergence of *C. felis*, at a given temperature there has been a proportion of the flea population observed by Silverman and Rust (1985) that remained in the cocoon for extended periods. Different emergence periods are thought to be caused by different environmental conditions and emergence stimuli, but under similar environmental conditions Silverman and Rust (1985) believe them to result at least partly from larval competition for food, with adults developed from light prepupae remaining for a shorter time in their cocoons. Water and food reserves dropping below a critical level, triggering emergence mechanisms, or a weaker cocoon produced by an incompletely nourished prepupa not impeding adult emergence could be possible reasons. But the primary factor responsible for initiating adult emergence and reducing the randomness of host location will be host-produced stimuli (Silverman and Rust 1985).

Pressure and heat are the two main stimuli inducing rapid emergence from the cocoon, in detail pressure of 13-254 $g/cm^2$ and temperature between 32-38°C (Silverman and Rust 1985). Since the combination of warmth and pressure provide higher emergence rates than either warmth or pressure alone, it is likely that an endothermic animal resting on a cocoon increases the chance that the adult flea would emerge and successfully attack the host (Silverman and Rust 1985). A man with a body weight of >75 kg walking over a carpet containing cocoons induces the emergence of 31% of the cocoons' population after the first walk, 97% after the fourth and all the imagines emerge after the fifth time (Silverman and Rust 1985). In the absence of stimuli adults emerging gradually over several weeks, depending on ambient temperature, with the length of time spent in the cocoon related to prepupal weight (Silverman and Rust 1985).

Frequent hosts of cat fleas such as domestic and feral cats and dogs, mustelids, and opossums do not necessarily return to flea-infested lairs, so that successful attack of a mobile host necessitates immediate emergence and host-seeking behavior (Silverman and Rust 1985), making the emergence stimulated by warmth and pressure understandable.

To summarize, the preemerged adult represents a developmental stage of the cat flea offering the possibility to survive non-parasitic periods without being harmed (Silverman and Rust 1985).

Summary see Box 4.

---

**BOX 4. Cat flea preemerged adults**

- **The "waiting stage"**
- **Emergence after 10 days or 6 months (known as "the pupal window")**
- **Survival stage of non-parasitic-periods (no host available)**
- **Important feature of evolution, due to coevolution with very mobile hosts**
- **Stimuli for rapid emergence are:**
  **- pressure (walking by a potential host)**
  **- heat (body temperature of a potential host)**

---

## Adults

Once the flea emerges from the cocoon, it will not undergo any further molts, and the only size increase occurs due to swelling of the abdomen after feeding (Dryden 1989a). To identify the different species of *Ctenocephalides* spp. the outer appearance of the imagines can be used (see also chapter 'General Morphology'). Key distinguishing features of the two most important species of pets, the cat and the dog flea, have been documented by Soulsby (1982):

*C. felis*: Both pronotal and genal combs are present; length of head is generally greater than twice the width; first two spines of the genal comb are approximately equal in length. The tibia of all 6 legs is armed with 4 to 5 teeth (see Fig. 2).

*C. canis*: Both pronotal and genal combs are present; length of head is not twice as wide; first spine of genal comb is noticeably shorter than the second spine. The tibia of all 6 legs is armed with 7 to 8 teeth (see Fig. 2).

After emerging from the cocoon, the flea almost immediately begins seeking a host (Dryden 1993) searching for a blood meal (Dryden 1989a). A variety of stimuli attract newly emerged fleas. Visual and thermal factors have been found to be primarily responsible for attraction and orientation to the host (Osbrink and Rust 1985b). Combinations of different stimuli including tactile stimuli, $CO_2$, air currents and light together with the adult's age stimulate locomotion and modify the adult's responsiveness, at the same time limiting environmental interference while locating the host (Osbrink and Rust 1985b). Stimuli that failed to elicit an observable response were substrate vibrations, cat odor, sounds, changes in light intensity, and the passage of shadows

(Osbrink and Rust 1985b). Visual stimuli have shown to be attractive to the cat flea, but even in their absence, fleas were attracted by heat with air currents (created by a warmed moving target), as well as $CO_2$ increased flea activity (also reported by Benton and Lee 1965), quantifiable only in the absence of visual stimuli (Osbrink and Rust 1985b).

Fleas possess specialized, powerful legs for jumping onto a host, and according to Osbrink and Rust (1985b) their jump seems to be directed but not precise, responding to the amount of stimulus and not to the pattern. Thirty-four centimeters have been recorded in jumping (Dryden 1996). An increase of the size of the visual stimulus increases the response of the stimulus, thus a potential host is the more attractive to the flea the larger its size is (Osbrink and Rust 1985b). Osbrink and Rust (1985b) could not observe any visual orientation under red light or an increase in attractiveness by increasing the complexity of patterns on the target, so that acute form vision does not occur in the cat flea (Rothschild and Clay 1952; Osbrink and Rust 1985b). Additional stimulation in form of air movements were necessary to evoke a directed jump onto a stationary heated target, which was simply causing attraction and orientation in the cat flea (Osbrink and Rust 1985b).

Adult cat fleas display positive phototaxis and negative geotaxis in both the unfed and engorged state (Dryden 1988). By that the newly emerged cat flea residing in the carpet will move on top of the carpet canopy where it will be able to jump onto a passing host (Dryden 1989a), enhancing the success in host acquisition (Dryden and Rust 1994). The cat flea has proved to be most sensitive to light with wavelengths between 510 and 550 nm (green light) and insensitive to wavelengths between 650 and 700 nm (Crum et al. 1974; Pickens et al. 1987). The flea's responsiveness to certain wavelengths of light explains observations that adult fleas congregate around vents to crawl spaces, entrances to dog houses, and window sills (Dryden and Rust 1994). Their responsiveness to light can be used to capture fleas in light traps (Dryden and Rust 1994) (see chapter 'Indoor Treatment').

The activity of cat fleas peaks at dusk (Koehler et al. 1989; Bossard 1997), which coincides with cat activity (Kern et al. 1992).

Newly emerged, unfed cat flea can survive several days before taking a blood meal. In cool, dry air, 10% of newly emerged cat fleas survived for 20 days, while in moisture-saturated air, 62% survived for 62 days (Silverman and Rust 1985). At 24°C and 78% RH, 95% died within 15 days (Dryden 1988), and under ambient room conditions averaging 22.5°C and 60% RH, 95% died within 12.3 days (Dryden 1989a). No life cycle stage (egg, larva, pupa or adult) can survive for ten days at 3°C or five days at -1°C (Silverman and Rust 1983). Survival rates of fed imagines without any given environmental conditions were stated as 234 days (Bacot 1914), 58 days (Soulsby 1968) and 11.8 days (Osbrink and Rust 1984). At temperatures of 5-15°C and 70-90% RH *C. canis* imagines were observed to survive seven days and *C. felis* adults an average of ten days. Derived from investigations with different environmental conditions, the survival time of unfed adult fleas increases with increasing humidity and sinking temperature and varies between 0.5 days at 35°C/2% RH and 40 days at 16°C/100% RH (Silverman et al. 1981b). Longevity of unfed adult cat fleas increases significantly in saturated air at 16°C compared to combinations of lower RH's and higher temperatures (Silverman et al. 1981b) (see Table 4).

For *C. canis*, adults were maintained at 8-10°C in saturated air for up to 58 days (Bacot 1914).

Low RH associated with subfreezing temperatures are likely to preclude adult cat flea survival off the host (Silverman et al. 1981b). During the warmer months of the year, it is doubtful that adult fleas can live for more than a week in the absence of a host or a suitable microclimate that is relatively cool and moist (Silverman et al. 1981b).

Once on a host, the cat flea begins feeding within seconds and mating occurs on the host in the first eight to 24 hours, with most females having mated by 34 hours (Akin 1984; Dryden 1990) (see also chapter 'Eggs'). Female cat fleas seem to have multiple matings, for young as well as fully mature and gravid females have been observed in the act of mating (Akin 1984). Furthermore it has been observed that the spermatheca (sperm holding organ) acquires progressively more sperm over the first 24 hours (Akin 1984). Multiple mating of one female with several males and sperm precedence which means that sperm deposited by the last male is the first used for fertilization, is often combined with protogony (i.e. females tend to develop before males) in insect species (Thornhill and Alcock 1983). Although most insect species exhibit protandry (males tend to emerge before females), cat fleas belong to a much smaller group that exhibits protogony (Thornhill and Alcock 1983). The multiple mating (Akin 1984) and the protogony may speak for possible sperm precedence in cat fleas (Dryden and Smith 1994).

After the first blood meal, the flea must continue to feed and reproduce in order to keep its metabolism in balance (Baker 1985). The adult flea is the perfect example of a parasite that must live on its host in order to survive. As an adult, its only function is to reproduce and it must feed constantly in order to do so (Baker 1985) (for egg production see chapter 'Eggs').

According to Zakson-Aiken et al. (1996) bloodfeeding is apparently necessary for oviposition as well as for successful mating. Males require feeding before the epithelial plug is unblocked in their testes (Akin 1984).

The importance of the blood meal for the larva has already been reported (see chapter 'Larvae').

Table 4. Effect of temperature and humidity on the longevity of unfed adult female cat fleas (*C. felis*) (average number of days for 90% mortality)

| Temperature (°C) | % relative humidity | | | | | |
|---|---|---|---|---|---|---|
| | 2 | 33 | 50 | 75 | 92 | 100 |
| 35 | 0.5 | 0.5 | 1.8 | 2.0 | 4.0 | 9.5 |
| 27 | 1.7 | 2.5 | 3.7 | 6.0 | 8.0 | 11.0 |
| 16 | 3.0 | 6.0 | 8.0 | 17.0 | 22.0 | 40.0 |

For blood intake, the suctorial mouth parts, well adapted to piercing and sucking from the skin are used. The host's epidermis is penetrated by the flea's maxillae. A tube, the epipharynx, enters the capillary vessels and draws up blood while saliva from the maxillae is deposited in the surrounding tissue (Lavoipierre and Hamachi 1961) (Fig. 10). Thus minimal damage to the skin is caused (Lavoipierre and Hamachi 1961). The saliva of the cat flea contains a substance that may soften and spread dermal tissue, assisting in the penetration of the dermis by the proboscis (Feingold and Benjamini 1961). It further contains an anticoagulant, helping in the uptake of blood (Deoras and Prasad 1967). The flea requires a period of between two and ten minutes to engorge (Rothschild 1975). The amount of blood consumed by a female cat flea is an average of 13.6µl (+/-2.7µl) per day, which is equivalent to 15.15-times the body weight (Dryden and Gaafar 1991). Female fleas increase their body weight by 40% during an one hour stay on a host while male fleas only show an addition of 3%. Within 48 hours the fleas reach their maximum weight. In females an addition of 140% (here 1.08 mg) respectively in males of 19% (here 0.43 mg) could be recorded (Schelhaas and Larson 1989).

Male fleas feed less frequently than females. Females were observed attached and feeding in one site for more than three hours, whereas males were rarely attached for periods longer than 10 to 20 minutes as reported for the bird flea (*Ceratophyllus idius*) (Schelhaas and Larson 1989). Unpublished laboratory observations by Pospischil (2001, personal communication) stated the maximum time of feeding of *C. felis* to be 5-10 minutes and of *Archaeopsylla erinacei*, the hedgehog flea, 20 minutes. Males do not only feed less than females, they are also more active on the host (Dryden 1990).

Once cat fleas feed on a host for a few days and initiate reproduction, they apparently reach a point at which they become dependent on a constant source of blood (Dryden 1993). By now the cat flea is thought to be a permanent parasite of its host. Dryden (1989b) found 85% of female and 58% of male cat fleas to be still present after 50 days on cats which have been restricted in their normal grooming activity (by declawing, fitting with an Elizabethan collar and housing in specially designed metabolic cages). Others report only a recovery of 22% of the fleas after 22 days on a cat (Hudson and Prince 1958). A permanent association of the cat flea with its host has already been described by Elbel (1951) and Deoras and Prasad (1967). Fleas leaving the host will either be dead or will die within four days (Dryden 1989a). Cat fleas which have fed for five days on a host and then been removed and held at approximately 24°C and 78%

Fig. 10. Adult cat flea feeding on human skin showing excretion of feces (blood)

RH, died within 48 hours (males) respectively 96 hours (females) (Dryden 1988). When fed only for twelve hours and then removed from the host, 5% were still alive at 14 days (Dryden 1993) so that once a few days are spent on a host a permanent and vital relationship seems to be established.

Maximum longevity of cat fleas has not completely been demonstrated, but survival on hosts which have been restricted in grooming activity has been reported for at least 133 days (58% of all the female fleas were recovered) (Dryden 1989b). Cat fleas housed in screen-covered microcells were reported to have an average on-host longevity of 7.2 days for males and 11.2 days for females (Osbrink and Rust 1984). *C. canis* has been reported to live for up to two years when fed on dogs (Harwood and James 1979).

An important role in the survival and longevity of fleas on the host is played by the grooming behavior of flea-infested animals (Hudson and Prince 1958; Osbrink and Rust 1984; Wade and Georgi 1988). Cats spend a considerable part of each day grooming themselves and have been shown to remove up to 50% of their flea parasite load within one week (Wade and Georgi 1988). Some pets may tolerate a small to moderate number of fleas, others groom themselves almost constantly, thereby ingesting and dislodging many of the fleas (Dryden 1993). Osbrink and Rust (1985a) reported of 70% of feline hosts having only relatively few (<7) fleas. Any cat flea dislodged from the host through grooming activity must return to the host or acquire another within a couple of days or the flea will die (Dryden 1993).

But there is also some form of interhost movement by the adult flea (Rust 1994), of importance also for epidemiology (see chapter 'Flea Epidemiology'). Infrequent, short-term contact between infested and uninfested hosts is insignificant for the movement of adult cat fleas (Blagburn and Hendrix 1989). But nevertheless movements between hosts of 2-15% of the cat fleas infesting a host are possible (Rust 1994).

Significantly more female cat fleas have been observed to remain on the host (Rust 1994). Movement occurs between hosts whether the hosts are permitted to live together or not. Movement by adult cat fleas between hosts occurs at a low rate and the likelihood of establishing new infestations by adult fleas transferring from one host to another exists, but does not seem to be as important as primary larval breeding sites (Rust 1994). Transference can also occur when cats and other hosts are killed and consumed as it is supposed for cat flea infestations of coyote, fox, and other carnivores (Rust 1994), and known to occur in weasels (*Mustela* sp.) (Marshall 1981).

Summary see Box 5.

**BOX 5. Cat flea adults**

- The final stage of insect metamorphosis

  Distinguished stage with females and males

- Host attack stimuli are: tactile stimuli, $CO_2$, air currents and light

- Unfed adults' survival time: about 20 – 62 days
  (dependent on climate)

- Fed adults on host survival time: up to 133 days

- Blood intake in female fleas: 13.6 µl/ day, equivalent to 15.2-times
  the body weight

- Start of feeding means regular blood meals necessary to survive

# Flea Epidemiology

Changing patterns in man's ecosystems caused by man have led among other things to a spreading of arthropod-related zoonoses in impoverished as well as in industrial countries. In Third World countries it is caused by a lower hygiene standard and the lack of hosts and reservoirs, and in industrial countries it is because of increasing cohabitation with domestic and exotic pets and the spread of synanthropic animals in urbanized areas (Rosiky 1978; WHO/WSAVA 1981; WHO 1988). Moreover, urbanization has acted as a selective factor which has led progressively to a drastic decline in biological diversity (Brown and Roughgarden 1989), while a restricted number of animal species has been demonstrated to be able to share the urban habitat with man, these species being dependent on human activities to maintain themselves (Genchi 1992).

Within the urban environment domestic animals are controlled by man, but synanthropic species can grow enormously, having a continuous and quite unlimited food supply from human activities and human refuse (Genchi 1992). In such changing patterns of the urban ecosystem, parasites have more opportunities to spread and to maintain themselves than other microorganisms, because of their better adaptability (Genchi 1992).

For the flea, its biology makes it a very efficacious vector of many pathogens (Genchi 1992). In towns the cat flea is mainly synanthropic and maintains its life cycle indoors feeding on pets (Genchi 1992). However, it must be emphasized that only about 5% of the flea population lives and feeds on the animals, the remaining 95% (eggs, larvae and pupae) are spread around indoor habitat. In buffered microclimate situations such as buildings in towns, cat flea populations grow throughout the year, showing unexpected large peaks caused by the sudden breakdown of nymphal diapause of a very large number of individuals when habitations are abandoned by domestic animals and their owners for longer or shorter times (Genchi 1992).

The reasons that the cat flea is such an extremely successful and ubiquitous parasite are the wide range of possible hosts and its status as a rather permanent parasite (Dryden 1993; Grant 1996).

The stimuli of emergence for the preemerged adults from the cocoon, temperature and pressure (Silverman and Rust 1985), as well as the stimuli provoking the adult, light, temperature, air currents and $CO_2$ (Osbrink and Rust 1985b) can be provided by a variety of hosts. Even though cat fleas have been reported to have reduced reproductive maturation e.g. when fed on calves (Williams 1993), they are found on a variety of wild and domestic animals including opossum, raccoon, red squirrel, rabbit, brown rat, short-tailed shrew, coyote, red and gray fox, bobcat, Florida panther, skunk, striped skunk, Townsend chipmunk, lynx, mongoose, ferret, lizards, and several rodent species (Amin 1976; Forrester et al. 1985; Fox et al. 1966; Geary 1959; Haas 1966; Hopkins 1980; Hopkins and Rothschild 1953; Hunter et al. 1979; Lewis et al. 1988; Morlan 1952; Rocket and Johnston 1988), domestic rabbits, horses, cattle, sheep, goats, jackals, poultry, and koalas (Blackmon and Nolan 1984; Dryden et al. 1993; Griffin et al. 1983; Halliwell 1982; Joseph 1985; Joseph et al. 1984; Taryannikov 1983; Williams 1993; Yeruham et al. 1982; Yeruham et al. 1989; Zimmerli 1982). Altogether the cat flea has been found on more than 50 hosts throughout the world (Hopkins and Rothschild 1953; Williams 1986).

Recurrences of fleas in temperate climates even after eradication efforts may be found because of the prolonged survival of preemerged adults and because of the lack of penetration of carpets by insecticides (Dryden and Rust 1994) when they occur within a few weeks or months after treatment. Whereas for yearly recurrences the extremely wide host range for *C. felis* with a variety of non-domestic animals as reservoir for reinfestations may be an explanation (Dryden and Rust 1994).

Interhost movement of adult cat fleas has been observed between cats (Rust 1994), and transference can also occur when cats and other hosts are killed and consumed (Rust 1994), as it was described in weasels (*Mustela* sp.) (Marshall 1981) (see also chapter 'Adults').

According to Baker (1985) another reason for adult fleas leaving their host either by chance or by choice, is an exceeding of the baseline of the flea population. Normally a stable population density established at any given exposure level is relatively constant, but varies greatly with season time and host. When infestation levels are extremely high, e.g. when adults emerge during spring and summer, the population can become excessive and fleas might leave their host during the first day before the population settles down (Baker 1985). So the likelihood of establishing new infestations by adult fleas transferring from one host to the other exists, and probably does not necessarily require physical contact (Rust 1994). This could explain how some indoor pets acquire fleas (Rust 1994). However, Rust (1994) suggests visitations to environments capable of supporting immature stages of fleas, the primary larval breeding sites, to be a more important source of flea infestations. These environmental conditions where second hosts live are presumed the most critical factor for a continued establishment of secondary infestations (Rust 1988).

Since all the life cycle stages of *C. felis* are susceptible to desiccation, only those eggs that fall into protected microhabitats hatch larvae that will ultimately develop into adults (Dryden 1989a). As stated by Byron (1987), suitable breeding sites are not widespread in homes but confined to specific sites. Areas that may be suitable for flea development in the house are (Dryden 1989a):

- pet's bedding
- thick shag carpet
- carpeted or dirt floor basements

Potentially favorable developmental sites outside occur where there is moist soil and shade:

- dog houses
- flower beds
- areas under bushes
- damp crawl spaces
- gardens
- any places where the flea-infested animal might rest during the heat of the day

Areas in the house such as wood or tile floors and well-traveled hallways are less likely to support development. Likewise, open areas of the lawn that are exposed to prolonged sunlight offer poor growth conditions (Dryden 1989a).

Summarizing, significantly more fleas are found in rooms where pets spend most of their time resting (Osbrink et al. 1986).

Opportunities of flea exchange are created by host movement and interaction (Marshall 1981). In this context the ranges of pets as well as wild life have to be considered. Cats can have home ranges in urban areas of <1 ha or up to 270 ha rurally, depending on cat density and the availability and distribution of food (Liberg and Sandell 1988). Urban dogs may have home ranges of 1.5-2.6 ha (Beck 1973). Wild animals are mobile, increasingly abundant in urban areas and often serve as alternative hosts for the cat flea (Bossard et al. 1998).

The abundance of adult cat fleas fluctuates with seasonal changes. The warm months of spring and summer give rise to the highest numbers, whereas few are found during the cold months of late fall and winter (Metzger and Rust 1997). Infestations of cat fleas consistently recur during the warm months of the year (Osbrink and Rust 1985a). No life stage of the cat flea can survive extended periods of subfreezing temperatures, and no reports of a diapausing stage exist (Silverman and Rust 1983). Furthermore cat flea populations are rarely detected on domestic hosts during winter months, but reinfestation of unknown origin are nevertheless common in spring and summer (Metzger and Rust 1997). Therefore two hypotheses have been proposed as possible overwintering strategies of cat fleas (Metzger and Rust 1997):

1. Feral mammals, whose territories extend into urban areas, harbor cat fleas all year and represent the source of reinfestation for domestic animals (Dryden and Rust 1994).
2. An unknown percentage of preemerged adults remain inside the cocoons for extended periods and emerge when conditions are favorable for immature development.

For the so-called 'winter fleas' these two hypotheses of either long-lived residents surviving through the fall and winter on a host or examples which emerge during these months as part of a small population are proposed (Metzger and Rust 1996).

Cat fleas associated with dogs and cats in cold climates are most likely adults surviving on untreated dogs and cats or small wild mammals. And as these raccoons or opossums e.g. pass through yards in the spring or set up nesting sites in crawl spaces or attics, the eggs laid by surviving female fleas drop off and subsequently develop into adults (Dryden 1993). Metzger and Rust (1997) assume that feral hosts (small- to medium-sized mammals) inhabit shelters which are protected from temperature extremes, so that theoretically populations of cat fleas could develop all year and that urban reinfestation could then occur either directly from feral animal shelters or from flea eggs and adults that might dislodge from feral hosts in favorable locations near domestic animals. But the high incidence of reinfestation favors additional sources (Metzger and Rust 1997). Inadequate insecticide treatment, (Osbrink et al. 1986; Rust and Reierson 1989) as well as preemerged adults inside the cocoons could be these additional sources (Metzger and Rust 1997). From the studies of Metzger and Rust (1997) it can be suggested that adult fleas are able to overwinter inside the cocoons. At 15.5°C, some fleas emerged as late as 155 days after egg deposition (Metzger and Rust 1997). Additionally, adults may survive on hosts for >100 days (Dryden 1989b). Combined with the additional time of immature development, which can take over three weeks at low temperatures, this could account for at least eight months in a year (Metzger and Rust 1997). In temperate latitudes where cat fleas are found, conditions are favorable for development of immatures for at least four months. Therefore survival during short winters could be spent as quiescent adults inside cocoons, whereas long winters might require a combination of survival as quiescent adult followed by survival on host animals (Metzger and Rust 1997).

Indoor and outdoor survival of immature cat fleas was tested in North-Central Florida and proved that in protected microhabitats (e.g. inside a doghouse or under a mobile home) cat fleas are able to complete their development outdoors, thus maintaining continuous populations throughout the year (Kern et al. 1999). Nevertheless low nighttime temperatures and low daytime relative humidities are probable causes for outdoor winter mortality of flea larvae (Kern et al. 1999). Inside buildings high mortality of larvae can be observed when cool outside air with a specific atmospheric moisture content enters the air handling system and is heated. Then the relative humidity decreases and the moisture deficit of the air increases, indicating why heated buildings are drier in winter. But also simple central heating or air conditioning both cause a drying effect (Kern et al. 1999) and thus influence flea development. During summer months, the high larval mortality in sun-exposed areas and structures that trap heat confirms the deleterious effect of high temperatures (>35°C) to cat fleas (Kern et al. 1999).

The survival and maintenance mechanisms of *C. felis*, important for epidemiological considerations can be summarized as follows (Rust and Dryden 1997):

1. The presence of adults on domestic and feral cats and dogs.
2. The presence of adults on urbanized small wild mammals (such as raccoons and opossums).
3. A delayed development of immature stages in freeze-protected underground dens of wildlife.
4. A delayed development of pupae and emergence of adults in the in-home environment.

# Flea-Related Dermatitis

Directly and indirectly the flea is probably the cause of more dermatological disorders in small animals than all other etiological agents together (Keep 1983).

Concerning the pathogenic effect of flea infestation, differentiations between the direct consequences of a flea bite, the allergic reactions of the host and the effect of blood consumption by the fleas have to be made (Kalvelage and Münster 1991).

On the one hand there is flea bite dermatitis and on the other hand there is flea bite hypersensitivity which according to Halliwell (1985a) can also be defined as flea allergy dermatitis (FAD). Newer research and diagnostic techniques suggest that most of the cases of dermatitis without the presence of allergic hypersensitivity reaction do have hypersensitivities and that dermatitis without allergy is a rare phenomenon (Halliwell 1983, 1986). Nevertheless several authors still differ between the two clinical situations of flea bite dermatitis and flea bite hypersensitivity (FAD).

Once the flea has found a host for the uptake of blood a few test bites have to be performed until a blood leading capillary vessel has been found (Karandikar and Munshi 1950). The direct contact with the host's skin as well as these bites cause an irritation of the skin, pruritus and an insignificant, transient invasion of lymphocytes in the direct proximity of the biting site (Gothe 1985). These local reactions are suggested to be principally similar in all hosts, whereas in the so far examined animal species differences exist concerning allergic reactions (Kalvelage and Münster 1991).

## Flea Bite Dermatitis

As mentioned above local reactions in form of irritation of the skin, pruritus and an insignificant, transient invasion of lymphocytes in the close area of the flea bite caused by direct contact of the skin with the ectoparasite and its bite (Gothe 1985) are supposed to be similar in all affected animals (Kalvelage and Münster 1991).

Flea bite dermatitis with its changes simply caused by the bite itself has to be distinguished from host reactions based on an allergic response to flea bites, even though both present pruritus as the main sign (Kalvelage and Münster 1991).

According to Torgeson and Breathnach (1996) flea bite dermatitis can be seen in any age or breed of dog or cat. The physical presence of fleas, both by means of their movement and the action of sucking blood can cause much distress which clinically manifests by pruritus, alopecia, and crustiness, due to self-inflicted trauma (Torgeson and Breathnach 1996). Furthermore pin prick hemorrhage sites caused by adult fleas biting the host may be evident. Commonly these lesions are found along the dorsum, particularly over the base of tail and along the lumbar spine (Torgeson and Breathnach 1996).

The flea bite is seen as a red spot with a halo varying in width and elevation. Papules as well as pustules and crusts can furthermore be observed. Continuous scratching and biting by the host cause moistening or even purulent dermatitis and, in chronic cases, thickening of the skin and hyperpigmentation. In immense infestation a regenerative anemia can develop (Scott 1978). Severe skin irritation produced by frequent bites of fleas can also induce secondary bacterial dermatoses (Keep 1983). This secondary infection is variable and usually classified as mild to moderate with the overall degree of pruritus (Torgeson and Breathnach 1996).

Flea-infested cats rarely demonstrate any significant degree of scratching or pruritus, however excessive grooming has been observed (Kwochka 1987; Kwochka and Bevier 1987). Asymptomatic flea-infested cats are of epidemiological importance in the discussion on effective flea control for multiple-pet households (Scheidt 1988).

Flea bite dermatitis is unquestionably the most common veterinary dermatological entity in the world (Halliwell 1979, 1983) and may be the most common overall disease problem of domestic animals (Halliwell 1983). In the southeastern part of the United States, flea-related diseases may account for over 50% of dermatological cases presented to the veterinarians (Halliwell 1979) and in excess of 35% of the total small animal veterinary effort (Halliwell 1985b).

## Flea Allergy Dermatitis (FAD)

Flea allergy dermatitis (FAD) is a disease in which a hypersensitive state is produced in a host in response to the injection of antigenic material from the salivary glands of fleas (Dryden and Blakemore 1989; Carlotti and Jacobs 2000). Synonyms for FAD include flea bite allergy (Muller et al. 1983) and flea bite hypersensitivity (Michaeli et al. 1965). In cats the disease is also known under the diagnosis of feline miliary dermatitis and feline eczema (Reedy 1975). The condition is found worldwide and represents the most common dermatological disease of dogs and a major cause of feline miliary dermatitis (Baker and O'Flanagan 1975; Halliwell 1979; Muller et al. 1983) as well as the most common allergic skin disease of dogs and cats (White and Ihrke 1983; Reedy and Miller 1989a, 1989b). The association of fleas with what was called 'summer eczema' was first reported by Kissileff in 1938.

It has been reported that FAD is most prevalent in the summer months, however in warm climates, where flea infestation may persist throughout the year, the problem may be non-seasonal. In northern temperate regions, the close association of pets and their fleas with human dwellings may also create conditions that permit a year-round problem (Dryden and Blakemore 1989). The condition is typically worse in the summer and fall in temperate climates (Kwochka 1987).

Hypersensitivity to flea bites is not only of importance to domestic pets but is also an important cause of the common skin disease in man, termed papular urticaria (Maunder 1984). Detailed investigations carried out on patients exposed to flea-infested pets have shown that the incidence of such reactions is quite high (Hewitt et al. 1971; Kieffer et al. 1979).

## Clinical Signs

The clinical signs of flea allergy in the dog and cat are the result of pruritus and self trauma (Moriello 1991). They can be quite variable, depending on the degree of sensitivity, the level of flea exposure, the stage of the disease (acute versus chronic) and the presence of concurrent primary (atopy) or secondary (bacterial folliculitis) skin disease (Halliwell 1979, 1983, 1984).

Summarized by Dryden and Blakemore (1989) the clinical signs associated with FAD are variable and depend upon the following:
- frequency of flea exposure
- duration of disease
- presence of secondary or other concurrent skin disease
- degree of hypersensitivity
- effects of previous or current treatment

The classic primary skin lesion in the hypersensitive dog or cat is a small pruritic, erythematous wheal (immediate hypersensitivity) noted at the site of the flea bite (Halliwell 1983; Kwochka 1987). The immediate reaction may resolve or progress, if a delayed reaction is also present, into a pruritic, erythematous crusted papule (delayed reaction) (Halliwell 1979, 1983; Kwochka 1987). Frequently, both erythematous wheals and papules go unnoticed by the owner, who may only observe pruritus with or without the presence of fleas (Nesbitt and Schmitz 1978). The dermatitis is typically confined to the dorsal lumbosacral area, forming a typical triangular image (Bourdeau 1983) or an inverted V pattern (Merchant 1990), further confined to the caudomedial thighs, ventral abdomen and flank (Baker 1974; Muller et al. 1983; Halliwell 1979; Halliwell 1983). Other commonly involved areas are the neck, the cranial aspects of the forearm, and the base of the ears (Nesbitt and Schmitz 1978) (see Fig. 11). In severely hypersensitive animals, generalized cutaneous signs may develop (Lorenz 1980).

With prolonged exposure to fleas, the acute primary lesions are rapidly replaced by chronic secondary lesions that result from repeated self trauma and pruritus (Scheidt 1988). Diffuse erythema, excoriations, and partial to complete alopecia replace the initial wheal or papule. Broken or stubbled hairs are commonly noted along the lower back and may be observed in association with hairballs in the cat or hair-impacted feces.

Excessive wearing of the anterior teeth, primarily the labial and facial surfaces of both the upper and lower incisors, has been observed in dogs with chronic flea allergy that are historically chewers or nibblers rather than scratchers (Kuder, personal communication, cited in Scheidt 1988). Secondary seborrhea with odor (Bourdeau 1983: like rancid butter) is common (MacDonald 1983). Acute moist dermatitis is seen on occasion (Kwochka 1987). Areas of acute serous dermatitis ('wet eczema') up to 10 cm in diameter may occur on any part of the body which can be reached by the tongue, teeth or feet (Baker 1977). Long-standing pruritus and self-trauma may lead to marked secondary changes of acanthosis, hyperkeratosis, lichenification, and hyperpigmentation (Kwochka 1987). In the sensitized dog, scratching results in renewed pruritus, and a viscous circle is created even though fleas are no longer present (Baker 1977). Secondary staphylococcal folliculitis has been reported to occur in 10% of cases of flea allergy dermatitis (Nesbitt and Schmitz 1978), but the actual incidence may be much higher (Kwochka 1987). Veith (1989) summarizes possible secondary changes into three main canine and one feline syndrome: (1) papulocrustous dermatitis, (2) pyotraumatic dermatitis, (3) bacterial folliculitis concerning the dog and (1) feline papulocrustous dermatitis in the case of the cat (for further information see Veith 1989).

As already mentioned FAD has been reported to be the most common cause of miliary dermatitis in the cat (Foil 1986; Scott 1980). Miliary dermatitis may initially develop around the neck and lower back or be generalized along the dorsum (Scheidt 1988). The periauricular area, ventral abdomen, and caudal and medial hind limbs can also be afflicted (Sosna and Medleau 1992) (see Fig. 12). The nonspecific, eczematous lesion of miliary dermatitis is not an actual flea bite, but rather a systemic cutaneous response of the flea-allergic cat. Pruritus is usually generalized and severe in feline flea allergy, especially when large flea numbers are present. Flea-allergic cats may lick excessively, chew, scratch, or pull out large tufts of hair. Further possible, less common presentations are facial pruritus, exfoliative dermatitis, and dorsal alopecia (Foil 1986; Scheidt 1987) in which the hair loss patterns may be consistent with psychogenic alopecia, and finally raised, red lesions of the eosinophilic granuloma complex (Sosna and Medleau 1992). (A very effective characterization of miliary dermatitis with crusted lesions of the size of millet seeds was reported by Veith (1989) who described the

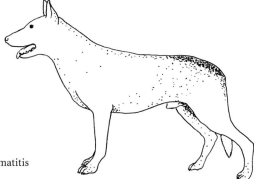

**Fig. 11.** Localization of flea allergy dermatitis on the dog

Fig. 12. Localization of flea allergy dermatitis
on the cat

posterior lumbar region as feeling as though covered with sand.) The classical triad of
pruritus, fleas and 'typical distribution' of clinical signs is not always present or
immediately apparent in the cat (Moriello and McMurdy 1989b). Pruritus is often masked
by over-zealous grooming instead of scratching (Moriello and McMurdy 1989b).

The variety of physical findings in the cat can be summarized, but not limited, to
the following (Moriello and McMurdy 1989b):

- localized or generalized miliary dermatitis (MD)
- eosinophilic plaques and/or ulcers
- bilaterally symmetrical alopecia
- ventral abdominal alopecia
- seborrhea
- dorsal lumbosacral hair loss
- generalized pruritus without obvious skin disease

and further (Scheidt 1988; Foil 1986):

- alopecia
- facial dermatitis
- exfoliative dermatitis
- 'racing stripe' or dorsal dermatitis

Other clinical findings that may be present in flea-infested as well as flea-allergic dogs
and cats include *Dipylidium caninum* (tapeworm) infestation, peripheral eosinophilia,
loss of weight or general condition, and blood loss anemia in young kittens, puppies or
debilitated adults (Nesbitt and Schmitz 1978; Scott 1980). Severely affected animals can
further be depressed and inappetant (Torgeson and Breathnach 1996).

For more detailed information on the clinic of FAD in dogs and cats see Scheidt (1988).

The most common age of onset for flea allergy dermatitis is three to six years,
although it may occasionally be seen in very old dogs and cats or those as young as six
months (Nesbitt and Schmitz 1978). In an evaluation of positive skin tests towards flea
allergen, it was found that the incidence peaked at two to three years and fell significantly
at six to seven years (Halliwell et al. 1987). Sex or breed predilection has generally not
been reported (Baker and O'Flanagan 1975; Muller and Kirk 1976a, 1976b; Nesbitt and
Schmitz 1978).

Table 5. Developmental sequence of flea saliva hypersensitivity in guinea pigs

| Stage | Time Period |
|---|---|
| I. Induction period | Lasts 4 days |
| II. Delayed reactions | 4 - 9 days |
| III. Delayed and immediate reactions | 9 - 50 days |
| IV. Immediate reactions only | 50 - 88 days |
| V. Non-reactivity | 88 - 140 days |

The development of flea hypersensitivity has also been studied in the in vivo guinea pig model, while the development of the disease was not comparable in all stages with the one in dogs or cats (see Table 5 and Chapter 'Immunopathogenesis of FAD').

For terminology of dermatitis and immunology related to flea allergic dermatitis see Box 6.

---

**BOX 6. Terminology**

| Terminology | Explanation |
|---|---|
| **Allergy** | Congenital or acquired specific changes of the immune system to react against molecules. Depending on the type of allergen and procedure of development of immunological and clinical changes different types of allergy are distinguished. Xenobiotic substances that usually are non pathogenic will react as allergens. |
| **Atopy** | Congenital disposition to develop a primarily humoral response to allergens. Allergens in this case are antigens that mainly induce the production of anaphylactic antibodies of the immunoglobulin class IgE. |
| **Erythema** | gr. erythma – redness, reddening<br>Redness of the skin. Inflammatory reaction of the skin, due to hyperemia. |
| **Excoriation** | Abrasion of the skin layer reaching the corium. |
| **Hapten** | gr. haptein – to adhere<br>Often low molecular substances. Haptens are not capable to induce an immune response on their own. In conjunction with a carrier they react specifically with antibodies or lymphocytes comparable to complete antigens (e.g. molecules such as penicillin can induce an immune response in conjunction with a carrier, here a protein). |

| Terminology | Explanation |
|---|---|
| **Hyperkeratosis** | Increased thickness of the stratum corneum of the skin either by increased production of keratinocytes (proliferative hyperkeratosis) or by a reduced rejection (retentive hyperkeratosis). Ectoparasitism suggests a parakeratotic, or nucleated, diffuse hyperkeratosis. |
| **Parakeratosis** | Increased thickness of the stratum corneum in conjunction with an incomplete maturation in epidemopoiesis. Cells with nuclei are still present, skin shows crusts and often secondary dermatitis. |
| **Pruritus** | lat. prurire – to itch syn. itching; Unpleasant sensation of the skin that provokes the desire to scratch. Common symptom in dermatology, localized or poorly localized. There are several mediators and modulators known to be involved in pruritus. |
| **Pyoderma** | Scab- or puss-eruption. Superficial or deep disorder of the skin and its appending structures (hair, nails, sweat glands), usually caused by staphylo- or streptococci. Classified as primary or secondary, superficial or deep. Most are superficial and secondary. Pyogenic infection of the skin, common in dogs but uncommon in cats. |
| **Seborrhea** | Increased and pathologically altered discharge of the sebaceous glands due to disposition. Usually a chronic disease of dogs, characterized by a defect in keratinization. Clinically resulting in increased scale formation, occasionally excessive greasiness of the skin and hair coat and sometimes secondary inflammation. |

## Differential Diagnosis of FAD

For detailed information on the differential diagnosis of flea allergy dermatitis in the dog and the cat see among others Scheidt (1988). In this context there is only a short tabular summary of the important differential skin diseases with the adherent diagnostic test to rule them out (Table 6, Table 7).

Table 6. Differential diagnosis for feline miliary dermatitis and diagnostic tests to identify specific cause

| Disease | Diagnostic test |
| --- | --- |
| Flea bite hypersensitivity | Intradermal skin test with aqueous flea antigen |
| Cheyletiella dermatitis | Skin scrapings<br>Acetate tape preparation |
| Pediculosis | Skin scrapings |
| Food allergy | Elimination diet |
| Atopy | Intradermal skin testing |
| Notoedric mange | Skin scrapings |
| Otodectic mange | Ear smear<br>Skin scrapings<br>Otoscopic examination |
| Trombiculiasis | Skin scrapings |
| Dermatophytosis | Fungal culture<br>Skin biopsy |
| Drug eruption | History of drug exposure<br>Skin biopsy |
| Bacterial folliculitis | Positive identification of pathogenic bacteria on skin cultures |
| Idiopathic miliary eczema | Rule out other causes |

Table 7. Differential diagnosis for FAD in the dog and diagnostic tests to identify specific cause

| Disease | Diagnostic test |
| --- | --- |
| Sarcoptic mange | Skin scrapings<br>Response to scabicidal dip |
| Food allergy | Response to elimination diet (2 to 3 weeks) |
| Allergic inhalant dermatitis | Positive reactions on intradermal skin test<br>Positive antigen-specific serum IgE levels |
| Bacterial hypersensitivity | Positive reactions to intradermal staphylococcal cell wall antigens<br>Response to antibiotic therapy |
| Intestinal parasite hypersensitivity | Positive fecal flotation<br>Response to anthelmintic therapy |
| Drug eruption | Skin biopsy and histologic findings<br>History of drug exposure |
| Demodicosis | Positive skin scrapings |

## Histopathological Findings

Biopsies of dogs with FAD showed the principal epidermal changes of hyperkeratosis and acanthosis (Nesbitt and Schmitz 1978). In some sections the same authors further found marked parakeratosis and surface scabs, but the most consistent change in the dermis was moderate to marked infiltration of lymphocytes, histiocytes, and plasma cells, often accompanied by small numbers of neutrophils and mast cells. Eosinophils were only observed in small numbers. Vascular dilatation ranged from nonexistent to marked (Nesbitt and Schmitz 1978). Baker (1977) reports of vascular dilation in the upper dermis in early lesions, with moderate acanthosis and upper dermal edema. In the acute serous stage, complete loss of the dermis and upper dermal edema with marked upper dermal inflammatory infiltration composed of neutrophils and predominating eosinophils have been present. In the chronic stage a markedly thickened epidermis, with extensive infiltration of the upper dermis by plasma cells, lymphocytes, neutrophils, predominating with eosinophils, and finally mast cells could be observed, all accompanied with vascular dilation as in the other stages (Baker 1977).

But in general, skin biopsies from flea-allergic animals are nonspecific and reveal varying degrees of eosinophilic perivascular dermatitis and epidermal edema (Kwochka and Bevier 1987). Similar findings are observed in biopsies of dogs with contact allergic dermatitis (Nesbitt and Schmitz 1977).

## Immunopathogenesis of FAD

The association between an ectoparasite and its host is quite variable (Nesbitt and Schmitz 1978). According to Gaafar (1966) the four basic reactions of a host to its ectoparasitic invader can be classified in one or more of the following reactions: (1) histaminic, (2) enzymatic, (3) hypersensitive or allergic, (4) immune and non-responsive.

Histamine and histamine-like compounds have been described as constituents of some salivary secretions and venoms of many arthropods. The appearance of a wheal at the invasion site within a few minutes suggests a histaminic reaction. The initial reaction to flea bites is seen in all animals, but a sustained reaction is seen only in those that are hypersensitive or immunologically competent (Nesbitt and Schmitz 1978).

An enzymatic reaction is mediated by the secretions of the parasite. It is used to dissolve parts of the tissues and to provide the parasite with an avenue to its predilection site. These secretions are made up of enzymes with proteolytic, anticoagulant, cytolytic, and spreading action. Digestion and liquefaction of the host tissue results in the production of substances foreign to the body, with subsequent local inflammatory reactions in the tissue. When enzymes in the secretions of the parasite are injected into the tissues of the host, they become antigenic as well as cytolytic. As a result of absorption of the antigen at the infestation site, immunologically competent animals may become sensitized, with a hypersensitivity state developing on repeated exposures. The immune and nonresponsive type of reaction is the result of immune tolerance characteristic of many parasitic infections, including FAD (Nesbitt and Schmitz 1978).

The immunopathogenesis of FAD in the dog is multifactorial and includes factors such as genetic predisposition to develop allergic disease (atopy), type of exposure (intermittent versus continual), and age of exposure to a constant flea load (Halliwell 1984; Halliwell et al. 1987).

A sensitized animal in general can show two types of reaction (Bourdeau 1983):

1. A humoral reaction (immediate hypersensitivity of type I), which appears fast (after a few minutes), caused by the interaction of antibodies and immunoglobulins of type E fixed on mast cells. A degranulation follows with release of a high quantity of

inflammatory mediators, especially histamine, serotonin and additionally diverse leucotrienes. Acute reaction, rapidly produced, characterized by the presence of numerous eosinophilic cells.

2. A delayed cell-mediated reaction (hypersensitivity of type IV), containing an interaction of T-lymphocytes, the multiplication of these cells, a cellular migration, the liberation of lymphokines and the infiltration by monocytes, lymphocytes and transitory polynucleic neutrophiles. (In the course of this reaction cellular lesions with a release of lysozymes and a liberation of inflammatory mediators develop, causing pruritus and different lesions in the animal.)

Detailed investigations concerning the immunopathogenesis of allergic reactions to flea bites were originally performed using a guinea pig model. Benjamini et al. (1961) differentiated five sequential phases in guinea pigs repeatedly exposed to fleas (see also Table 5):

**Stage 1:** Induction of hypersensitivity with no observable skin reactions from day 1 to 4.

The 1st stage or induction period coincides with the first flea bite, which constitutes the initial antigenic stimulus for the synthesis of antibodies. During this stage, neither humoral nor cellular antibodies are demonstrable (Feingold et al. 1968).

**Stage 2:** Delayed skin reactions from days 5 to 9.

The 2nd stage of delayed reactivity, which manifests a reaction twelve to 24 hours after exposure, is characterized macroscopically by a papular eruption, erythema and induration and microscopically by infiltration of lymphocytes, monocytes and plasma cells. These are all cell-mediated immune reactions (Feingold et al. 1968).

**Stage 3:** Immediate skin reactions, followed by delayed reactions from days 10 to 60.

Two to five days after the delayed reaction the guinea pigs respond to flea bites within one hour with erythema and edema, which fade after four to six hours. In most cases non-reactive spots also gain activity during this time. Then for about seven weeks immediate and delayed response appear together.

The general mechanism of immediate hypersensitivity in the 3rd and 4th stage involve humoral antibodies. These may be mediated by reagenic antibodies (IgE) or by cytotoxic antibodies and complement (Feingold et al. 1968).

**Stage 4:** Immediate reactions only, from days 61 to 90.

**Stage 5:** No skin reactivity after 90 days.

The animal does not respond to exhibition of fleas anymore and has become tolerant.

Without a repeated exposition a hypersensitive state once developed will persist for at least twelve to 18 months (in guinea pigs; Benjamini et al. 1961). Experiments with dogs have not indicated the same orderly progression of reactions as in guinea pigs (Gross and Halliwell 1985). Intermittent exposure of flea naive dogs to fleas resulted in no consistent pattern of development of hypersensitivity in the early sensitization period. Immediate reactions prior, later or at the same time could be observed, with not all dogs having the same sequential pattern of cutaneous response (Gross and Halliwell 1985). According to Bourdeau (1983) the 2nd and 3rd stages are the most frequent in

FAD of the dog, and theoretically one flea is sufficient to provoke them. Apart from the missing sequential pattern in dogs, this animal species displayed continued reactivity throughout a 44-week study when exposed intermittently to flea bites (Gross and Halliwell 1985; Halliwell 1984), in contrast to guinea pigs, reaching a state of non-reactivity after 20 weeks of repeated exposure to flea bites (Benjamini et al. 1961). Nevertheless epidemiologic studies imply that the degree of sensitivity may lessen over years (Halliwell 1984). For detailed reactions in the dog see Table 8.

The immunopathogenesis of FAD is highly complex. According to Halliwell et al. (1987) in the dog FAD is mediated by at least three distinct immunopathogenic pathways: (1) immediate hypersensitivity, (2) cutaneous basophilic hypersensitivity (CBH) and (3) delayed hypersensitivity. In following studies, Halliwell (1990) also postulates the possible participation of an Arthus reaction in the immunopathogenesis of FAD.

Before the uptake of blood, the flea excretes saliva into the wound (Dryden 1989a). The saliva of *C. felis* contains a substance that may soften and spread dermal tissue, thus assisting in the penetration of the dermis by the proboscis (Feingold and Benjamini 1961). The flea saliva further contains an anticoagulant that assists in the uptake of blood (Deoras and Prasad 1967) and also has important antigenic properties.

In inducing flea bite hypersensitivity major emphasis was placed on an incomplete antigen, or rather hapten of relatively low molecular weight which has proven to be stable to heat (Benjamini et al. 1963). This flea hapten is associated with both salt-soluble and acid-soluble fractions of collagen, which act as a carrier (Michaeli et al. 1965). The hapten is not antigenic on its own, but becomes a complete antigen upon union with dermal collagen, being responsible for the cell-mediated allergy type IV.

According to Halliwell (1984) an immediate wheal following flea bite or injection of allergen could not occur because of this hapten alone, as the release of histamine from mast cells, which generates the wheal, results from the bridging of two adjacent molecules of IgE antibodies on the mast-cell surface by a bivalent antigen, such causing degranulation of the cell. This cannot be accomplished physically by a simple chemical

Table 8. Phases of flea allergy dermatitis in dogs

| Phase | Stage | Type of reaction/Occurrence | Clinical diagnosis |
|---|---|---|---|
| Preclinical | Stage I | Sensibilisation to an allergen No characteristic cutaneous lesions | Intradermal reaction negative |
| Clinical | Stage II | Retard hypersensibility High frequency | Intradermal reaction positive within 48 hours |
| | Stage III | Immediate or retard hypersensibility Frequent | Intradermal reaction often positive |
| | Stage IV | Immediate hypersensibility | Intradermal reaction positive ( 20 minutes) Desensibilisation effective |
| Desensitization | Stage V | Desensibilisation rare | Intradermal reaction negative |

of low molecular weight (Halliwell 1984). Apart from the low-molecular weight hapten, Halliwell (1986, 1987) proposes at least two additional allergens of more than 20,000 dalton molecular weight as allergenic components of flea saliva for the immediate allergic reaction (reviewed by Kwochka 1987).

In flea-allergic dogs significant levels of IgE as well as IgG antibodies were assayed in contrast to low levels in flea naive dogs or non-allergic dogs, chronically exposed to fleas (Gross and Halliwell 1985; Halliwell and Longino 1985). In the immediate allergic reaction (type I, IgE mediated) reagenic antibodies react with sensitized mast cells to elicit an inflammatory response that is predominantly eosinophilic (Gross and Halliwell 1985). The delayed response may represent in part a cell-mediated allergic response (type IV), characterized by an influx of T-cells which have been sensitized to allergen presented in lymph nodes by cells of the mononuclear phagocyte series (Gross and Halliwell 1985). Furthermore possible involvement of a late-onset IgE-mediated response in conjunction with the classic immediate hypersensitivity reaction and alternatively of cutaneous basophilic hypersensitivity (CBH) is suggested and may contribute to the development of flea allergy in the dog (Gross and Halliwell 1985; Halliwell and Schemmer 1987). Clinically late-onset IgE-reactions are similar to type I reactions except that they occur approximately two to six hours after the insult and not within 15 to 30 minutes post-insult (Moriello 1991).

Predisposing or influencing factors for the development of flea allergic reactions are the scheme of exposure to fleas, the age of exposure and underlying diseases such as atopy respectively allergic inhalant dermatitis (Halliwell et al. 1987).

According to Halliwell (1984) all animals can become allergic to fleas if they are appropriately exposed. Intermittent exposure favors the development of flea allergy (Gross and Halliwell 1985; Halliwell et al. 1987), whereas continuous, chronic exposure retards the development of allergy, which is then less severe if it does develop, suggesting that this scheme of exposure may result in partial or complete tolerance (rather than hyposensitization in the common sense) (Halliwell and Longino 1985; Halliwell et al. 1987). Additionally, if flea-allergic dogs are left untreated, the sensitivity will gradually disappear (Halliwell 1985c). Finally, differences in flea burden in freely mingling dogs (Fehrer and Halliwell 1987) may imply some unknown resistance in some animals making them less attractive to fleas (Moriello 1991).

The age at which exposure first occurs may be relevant, with early exposure possibly protecting against the later development of clinical flea allergy dermatitis (Halliwell et al. 1987).

Last but not least a higher incidence of dogs with allergic inhalant dermatitis (atopy) has been shown to be allergic to fleas, indicating a possible inherited predisposition to develop FAD (Halliwell 1984; Nesbitt 1978). The possibility exists that such animals could be more readily sensitized to other allergens via other routes (Halliwell et al. 1987).

Although the pathophysiology of flea allergy in the cat has not been completely identified as it has in the dog, similar pathomechanisms are felt to exist (Foil 1986), but only type I reactions have been demonstrated in cats so far (Moriello 1991).

Cross reactivity between bites of different flea species has been demonstrated in guinea pigs. Guinea pigs sensitized to bites of *C. felis* also reacted to the bites of *P. irritans* and

*P. simulans* thus indicating the possible presence of a common antigenic factor in all three flea species (Hudson et al. 1960). Baker and Mulcahy (1986) showed that the intradermal injection of dogs with either flea saliva or whole flea extract from *C. canis* or *A. erinacei* produced similar immediate hypersensitivity reactions. Similarly, a rabbit flea, *Spilopsyllus cuniculi*, can cause hypersensitivity reactions in the cat (Studdert and Arundel 1988).

## Diagnosis of FAD and Flea Bite Dermatitis

Most diagnoses of FAD can be drawn upon the following basics (Nesbitt and Schmitz 1978):
- history
- distribution of lesions
- elimination of other allergic and non-allergic diseases
- evaluation of the response to treatment

A number of diagnostic procedures can be used to confirm the diagnosis, and can be performed according to indicated need, as there are (Nesbitt and Schmitz 1978): skin scrapings, fungal and bacterial cultures, complete blood counts, intradermal testing, isolation from suspected contact allergens and subsequent provocative exposure to them (Nesbitt and Schmitz 1977), diet testing, and finally biopsies for histologic studies (Allen and McKeever 1974) (see also chapter 'Differential Diagnosis of FAD').

Integrating the historical and physical findings with the demonstration of fleas or flea excreta usually makes the diagnosis of flea bite dermatitis or flea allergy in the dog or cat quite obvious (Scheidt 1988). In fact, fastidious grooming may remove fleas and their excreta (Kwochka 1987) just as recent bathing or grooming of the pet prior to examination can account for the lack of fleas on the pet (Scheidt 1988). This and the fact that many owners have difficulty in accepting that socially unacceptable creatures like fleas are responsible for their pet's skin lesions (Baker 1977) should be considered when examining a pet for fleas. The use of a fine-toothed metal comb, a flea comb, can help to pick fleas out of a coat or aid in harvesting particles of flea excreta (Kissileff 1969). Additionally, visualization of debris left on the examination table with a hand-held magnifying lens may reveal the ovoid (0.5 mm), white eggs of fleas (Kwochka 1987). The flea comb can be used either alone or following application of an insecticidal spray, as demonstrative method for the owner (Kissileff 1969). The fact that flea excreta are dried blood components (see chapter 'Larvae') is a very impressive evidence of prior flea invasion by moistening the flea excreta with water on a white paper towel or cotton ball and observing the characteristic red-brown color (Kissileff 1969). To achieve the evidence of fleas on the host or the host's bedding and habitat a vacuum cleaner with fine gauze inserted behind the nozzle can also be used and applied to the host and the host's habitat. Fleas are retained on the gauze afterwards (Urquhart et al. 1987).

Further diagnostic efforts for a positive flea diagnosis can be a fecal examination of the animal for the presence of *D. caninum* segments or eggs, for which the flea can act as intermediate host. In a small number of cases, hematology may reveal a regenerative anemia and peripheral eosinophilia, and biopsy may show a perivascular dermatitis with eosinophils (Baker 1974; Baker and O'Flanagan 1975; Kieffer and Kristensen 1979; Kristensen and Kieffer 1978; Scott 1980). An evaluation of the response to an adequate flea control program is the most widely used method of diagnosis (Nesbitt 1983), but since a single flea may elicit a severe pruritic episode in a hypersensitive host (Baker and Thomsett 1990), this method must be considered to have limited value (Stolper and Opdebeeck 1994).

Halliwell and Longino (1985) reported of high flea antigen-specific IgE levels in dogs with clinical FAD that could also be associated with positive skin responses in intradermal skin tests in studies by Hickey et al. (1993).

The final possible diagnostic tool for the verification of a flea allergic problem is the intradermal skin test (IDST). Besides the documentation of a hypersensitive state it is also an important and excellent tool for client education (Scheidt 1988). According to Chamberlain and Baker (1974) the IDST is the most reliable and reproducible skin test to perform. A 1:1000 (weight per volume) aqueous solution of antigen extract (Flea Extract 1:100, Greer Laboratories, Lenoir, North Carolina) is suggested for diagnostic purposes in the demonstration of hypersensitivity (Halliwell 1979, 1983, 1986) and appears to be useful in both dogs and cats (Halliwell 1979, 1983, 1986; Kunkle and Milcarsky 1985). An intradermal skin test is a simple immunologic test for determining the presence of cell bound reagenic skin sensitizing antibodies (Moriello and McMurdy 1989b).

**IDST** (summarized by Kwochka 1987):

The test side on the lateral thorax is gently shaved and cleaned and then followed by intradermal injection of 0.05 to 0.1 ml of the 1:1000 (weight per volume) aqueous solution. Equal quantities of diluent (usually phosphate buffered saline) and a 1:100,000 (weight per volume) solution of histamine phosphate are used as negative and positive controls, respectively. A positive immediate hypersensitivity reaction will produce a cutaneous erythematous wheal within 15 minutes. The site of flea antigen injection should be reexamined at 24 and 48 hours for a delayed reaction (Halliwell 1984). The delayed reaction may not occur as a discrete wheal, but rather very subtle, manifesting only as a diffuse edema with no recognizable papule (Kwochka 1987).

The evaluation of the IDST is performed by comparing the size and appearance of the wheal at the antigen injection site with the size and appearance of the negative control (subtraction from one another) (Nesbitt and Schmitz 1978). For an optimal evaluation system it is strongly recommended that research workers and practitioners use a caliper ruler for measuring skin responses to allergens and take a differential of at least 5 mm between the test and the control injection as evidence of a positive response (Stolper and Opdebeeck 1994).

A positive IDST first clearly indicates that the animal has been exposed to fleas and second is highly supportive of FAD, though not diagnostic (Moriello and McMurdy 1989a). A positive IDST to allergens is proof of the presence of skin-sensitizing antibodies, but does not necessarily mean that these antibodies are relevant to the pathogenesis of the disease (Deboer 1989). Likewise, a negative reaction to flea antigen, either immediate or delayed, should not eliminate FAD from the differential diagnosis when clinical signs of flea-bite hypersensitivity are present (Baker and O'Flanagan 1975; Baker 1984). False-negative reactions can develop in case of recent treatment with corticosteroids, antihistamines, or tranquilizers or if the animal has been frightened (Kwochka 1987; Muller et al. 1983; Reedy 1977) as well as in case of poor techniques in form of subcutaneous injections (Moriello and McMurdy 1989a). As a general rule, systemic glucocorticoids, often used in treatment of FAD, should be discontinued a minimum of two to four weeks prior to skin testing or one week for each month of continual steroid therapy (Halliwell 1983; Moriello 1987).

Skin test reactions are best evaluated in a darkened room via trans-illumination with a strong light source from the side (Moriello and McMurdy 1989a).

Controversial discussion on the effectiveness of the IDST as a diagnostic tool in the demonstration of FAD exist (Baker 1971; Kieffer and Kristensen 1979; Kristensen 1978; Kristensen and Kieffer 1978; Kunkle and Milcarsky 1985; Nesbitt and Schmitz 1978; Van Winkle 1981). Using commercial antigen MacDonald (1983) claimed to detect 90% of FAD cases, whereas Reedy (1986) reported of 45% of 193 suspected cases of FAD that reacted to commercial antigen. In a study by Slacek and Opdebeeck (1993) the use of a soluble antigen (FS), prepared from unfed fleas was most effective and detected 67% of the dogs reacting positively to feeding fleas and 57% of the dogs that had clinical signs. According to them the provocative test remains the most accurate test for identifying animals with FAD. Sèrtic (1965) reports of 87% positive reaction in dogs with clinical signs of FAD and Baker (1971) of nearly 81% positive reaction. Stolper and Opdebeeck (1994) identified 94% of dogs which reacted to feeding fleas using a soluble antigen (FS) and the condition of at least 5 mm difference in the mean diameter between antigen injection site and negative control at 15 and/or 30 minutes post application. After evaluating several studies Kalvelage and Münster (1991) suggest that the IDST with flea allergen is not very effective, as about 20% of flea allergy induced dermatitis cases are not registered.

In flea-allergic cats with miliary dermatitis most investigators found intradermal skin testing with flea antigen to be less specific and sensitive (Kristensen and Kieffer 1978; Kunkle and Milcarsky 1985), whereas Slacek and Opdebeeck (1993) detected 100% of clinically positive cats with a soluble antigen as well as 100% of the cats reacting positively to feeding fleas.

The problem presented to the practitioner whether the skin condition of a patient is due to FAD alone or also to other concurrent pruritic skin diseases such as scabies, food allergy, or allergic inhalant dermatitis (Muller et al. 1983), has to be solved by a thorough recording of the patient's history, clinical examination and different diagnostic methods (see also chapter 'Differential Diagnosis of FAD').

# Treatment of Flea-Related Skin Diseases

In this section only the principles of treatment in case of flea-related skin diseases will be outlined. The most important factor for a successful management will be an effective flea control program. For only by eradication of fleas a long-lasting successful treatment can be achieved.

The major cause of failure to eradicate fleas on a pet is a combination of poor organization and deficient client education (Scheidt 1988).

A three-pronged flea control program is advisable, consisting of (Kwochka 1987):

- treatment of the pet
- treatment of all contact animals
- treatment of the environment (indoors and outdoors)

The program then has to be individualized based on the number of animals in the environment, the size of the internal environment, the degree of outdoor exposure to fleas, the size of the outdoor area, the time of year, and the family situation, such as the presence of young children (Kwochka 1987).

The most important aspects of flea control for the owner to understand are that animals and environment must be treated at the same time and that continued treatment at regular intervals will probably be necessary. They must understand that although complete, long-term eradication may not be possible, good control with existing products usually is feasible (Kwochka,1987).

The basic control program again has three objectives (Nesbitt and Schmitz 1978):

- break the flea life cycle in the indoor environment and the kennel
- minimize flea infestation of all animals in the household
- control the allergic reaction of the flea bite

The first objective will be dealt with in more detail in the chapter 'Treatment of the Premises'.

The second objective, flea control and reduction of flea infestation on the pet, can be achieved and maintained in two ways:

1. The first is the chemical way of treatment with numerous sorts of insecticides. A vast number of active ingredients is available for flea control on the pet as well as in the environment. Here the aspects of a cumulative effect and potential toxicity are important to consider and can be avoided by no concurrent usage of the same chemical ingredient in the environment and on the pet (Kwochka 1987).

   Detailed information on insecticides and methods of application can be found among others in Kwochka (1987), Kwochka and Bevier (1987), MacDonald and Miller (1986), Kalvelage and Münster (1991), and Schick and Schick (1986), as well as in the chapters 'Biological Profile, Mode of Action (and Direct Insecticidal Activity)' and 'Imidacloprid as Veterinary Drug for Flea Control'. The success of the flea-control program will not only be determined by selecting the proper insecticide, but also by choosing a delivery system that is efficacious and fits the animal's personality and the owner's schedule (Kwochka 1987). A résumé on the different galenic formulations, e.g. dips, powders or dusts, sprays and foams, repellents is given in Kwochka (1987).

2. The second way is a more mechanical approach to flea control with an extensive fur care (Kalvelage and Münster 1991) which can especially be achieved by the regular use of a flea comb. The flea comb can also be used as a litmus test for an environmental control program and as a good tool to demonstrate to the owner the extent of his pet's flea problem (Shanley et al. 1992). The use of a flea comb is furthermore advisable if a pet is too young or too sick to have insecticidal preparations applied (Torgeson and Breathnach 1996).

   The third objective, the control of the allergic reactions, is dealt with in the following chapter.

## Adjunctive Medical Therapy

The scheme of treatment for the third objective (see above), namely the control of the allergic reactions, is dependent on the clinical presentation of the flea induced dermatitis. Supportive medical therapy must be instituted in order to control pruritus and secondary skin disease. In the hypersensitive animal, systemic glucocorticosteroid therapy is often needed to control inflammation and associated pruritus (Muller et al. 1983; Halliwell 1979; Kristensen and Kieffer 1978; Scheidt 1988). But it must be remembered that this therapy only alleviates the pruritus and does not cure the allergy (Dryden and Blakemore 1989). Corticosteroids will mask the response to flea control therapy and delay the identification of a definitive underlying cause for the problem (Moriello and McMurdy 1989a).

Antihistamines have been used alone or in combination with glucocorticosteroids to help control inflammation and pruritus (Dryden and Blakemore 1989; Scheidt 1988).

Systemic antibiotics in case of secondary pyoderma are very beneficial and frequently indicated (Scheidt 1988). An effective antibiotic should be selected based on bacterial culture and sensitivity results (Scheidt 1988) and normally requires a time schedule of ten to 21 days or longer (Torgeson and Breathnach 1996). Elimination of a concurrent secondary pyoderma will help to lower the level of pruritus and decrease erythema and odor in the flea allergic patient (Scheidt 1988).

Antiseborrheic shampoos and emollient rinses should be used in conjunction with topical flea control to achieve a removal of scale and an improvement of the skin quality (Scheidt 1988). (More detailed information can be found in Veith 1989.)

## Hyposensitization

The aim of desensitization therapy is to bring the animal to a desensitized stage as soon as possible, thus avoiding the foregoing reactive stages (Keep 1983). Hyposensitization consists of administering allergens to a hypersensitive animal on a regular basis in an attempt to obtain a state of clinical non-reactivity to the bite of fleas (Dryden and Blakemore 1989). In short the practice is based on the fact that when an injection or a series of injections of an allergen are given, antibodies of the IgG type, the so-called blocking antibodies, are produced. These antibodies circulate and in general do not fix to tissue. Thus they have the effect of binding any allergen before it can reach the sites of fixation of IgE (Halliwell and Schwartzman 1971).

Double-blind desensitizing studies in dogs and cats have shown little efficacy (Kunkle and Milcarsky 1985; Halliwell 1981). Nesbitt and Schmitz (1978) found generally good response to treatment in groups of dogs without flea hyposensitization but continuous flea control and maintenance of environmental flea infestation. About 40% of the 70 dogs that had not responded to initial flea control program without flea bite hyposensitization responded better after hyposensitization was initiated. But the duration of effect was relatively short, usually lasting from three to six months. Because of the complexity of the immunopathogenesis of FAD (see chapter 'Immuno-pathogenesis of FAD') it might be predicted that desensitizing approaches would not be very successful, as for the fact that only IgE-mediated allergic diseases are amenable to therapy with hyposensitization, which does not exclusively seem to be the case in FAD (Halliwell et al. 1987). Approaches aiming at inducing tolerance might be more fruitful in FAD and there might be a potential in the field of either prophylactic or therapeutic induction of tolerance employing conjugates of flea allergen with known tolerogens (such as polyethylene glycol or the copolymer of D-glutamic acid/lysin) according to Halliwell et al. (1987). Studies using alum-precipitated flea-antigens have shown that these preparations will produce marked antibody responses in dogs and may also potentially be of benefit (Schemmer and Halliwell 1987).

The effectiveness of hyposensitization in dogs and cats is controversial (Halliwell 1983; Kunkle and Milcarsky 1985; Lorenz 1980; Nesbitt and Schmitz 1978; Reedy 1975). In summary, according to Scheidt (1988) flea hyposensitization, by itself, has not been found to be an effective method of therapy for canine or feline FAD and may be reserved for use in flea-allergic animals that are not responsive to systemic glucocorticoids or that manifest severe side effects to glucocorticoid therapy.

# Treatment of the Premises

## Indoor Treatment

Besides the treatment of the infested as well as all other pets of the household a synchronized in- and outdoor treatment is of importance for a proper management of FAD. Furthermore the fact that only a small part of the flea life cycle is spent on the host has to be remembered in this context (see chapter 'Flea Epidemiology'). An equal or greater effort must be directed toward the environment (Kwochka 1987).

An effective treatment can include thorough sanitation and insecticide application that provides both instant and residual killing activity in the house (Fadok 1984; Reese 1981). Indoor sanitation will physically remove flea eggs and larvae from the environment (Kwochka 1987). Repeated vacuuming of the carpets, furniture, floors and baseboards prior to insecticidal therapy should be performed (Scheidt 1988). Vacuuming with a beater-bar type vacuum will remove about 90% of the eggs and 50% of the larvae from carpets (Olsen 1982). In another study, a beater-bar type vacuum cleaner removed 15-27% of the larvae, 32-59% of the eggs, and some of the larval food, with its effectiveness decreasing as the density of the carpet pile increases (Byron and Robinson 1986). Vacuuming will also remove 95% of emerged adult fleas and stimulate preemerged adults to emerge (Osbrink et al. 1986). The vacuum bag should be deposited after use and either be frozen or burnt (Kalvelage and Münster 1991). Flea powder or pieces of flea collars placed into the bag can aid in killing the flea life stages and have been advocated, but concerns about nonlabeled usage and aerosolization to toxic levels in the house have been discussed (Kissileff 1987). Floors should be mopped, with particular attention paid to cracks and crevices where organic debris and flea eggs accumulate (Fadok 1984). Steam cleaning of the carpeting is a highly effective method in severe infestation (Fadok 1984; Reese 1981). Animal and human bedding frequented by pets should be washed at 60°C for ten minutes in a washing machine to remove flea life cycle stages or washed and dried at the maximal heat setting to ensure flea death (Fadok 1984; Schick and Schick 1986). The environmental clean-up should be finished before insecticide use begins (Kwochka 1987).

Special attention should always be given in households with young children and crawling infants when deciding on insecticidal treatment.

For the insecticidal treatment numerous products are available. The most common methods include hand-held sprayers, foggers, or professional exterminators (Scheidt 1988). For further information on ingredients, usage and time schedule see e.g. Kwochka (1987), Scheidt (1988) and Schick and Schick (1986).

Apart from chemical treatment, the responsiveness of adult cat fleas to light has been the basis for the use of light traps to control fleas (Rust and Dryden 1997). The cat fleas will orient and move up to 8.4 m towards a light source, although their jumping and collection in traps are greatly enhanced by a short interruption of the light, which may mimic host shadowing (Dryden and Broce 1993). *C. felis* is most sensitive to light with wavelengths between 510 and 550 nm (green light) and insensitive to wavelengths between 650 and 700 nm (Crum et al. 1974; Pickens et al. 1987). A green-yellow filter with wavelength centered at 515 nm was found to be more than twice as attractive to adult *C. felis* as white light (Dryden and Broce 1993). The flea's responsiveness to certain wavelengths of light clearly explains observations that adult fleas congregate around vents to crawl spaces, entrances to dog houses, and window sills (Dryden and Rust 1994). The age of the flea affects its responsiveness to visual targets; 5- to 6-day-old adults are the most responsive (Osbrink and Rust 1985b). Fleas <24 hours old are less responsive to light traps than 1- to 3-day-old fleas, suggesting that some maturation period is necessary before fleas can respond to host stimuli (Dryden and Broce 1993). While the cat flea will orient and move towards a light source, its taxis is (as mentioned) greatly enhanced when the light source is suddenly and temporarily interrupted (Dryden and Broce 1993). Locomotion increases during the last two hours of the photophase and declines throughout the scotophase (Koehler et al. 1989). A light trap with intermittent pulses of light attracts 57% of the adult fleas over a distance of 8.4 m (Dryden and Broce 1993), a sevenfold increase over commercially available traps with continuous light sources (Rust and Dryden 1997). The same trap design can collect >86% of the fleas in a carpeted room (3.1 by 3.3 m) during a 20 hours trapping period (Dryden and Broce 1993). Generally, it is suggested that intervention with trapping as well as vacuuming may greatly reduce the need for repetitive chemical treatments and the likelihood of insecticide resistance (Rust and Dryden 1997). Light traps have further also been used in field studies with chemical compounds (e.g. Dryden et al. 1999b, see chapter 'Field Studies').

## Outdoor Treatment

The amount of flea control in the external environment will depend on the size of the area to which the animal is confined (Kwochka 1987). Smaller areas of confinement and parts of larger yards where the pet spends the majority of its time can be treated (Fadok 1984). Outdoor sanitation includes removal of as much organic material as possible by mowing, raking, and discarding debris (Schick and Schick 1986). In this context, special attention should be given to any favorite sleeping areas, such as under a cool porch in summer (Kwochka 1987) or inside the dog house. Alternative biological control has been reported in form of ants which are known to be predators of all stages of fleas except cocoons (Fox and Garcia-Moll 1961; Silverman and Appel 1984). The beetles Histerid, Staphylinid, and Tenebrionid have also been reported to prey fleas (Fox and Bayona 1968; Jenkins 1964). Finally the entomophilic nematode *Steinernema carpocapsae* attacks and kills larvae, prepupae, and pupae. But a complete nematode development is not supported by fleas, and moist substrates are necessary to facilitate a nematode survival and thus an infection of more than one generation of fleas (Silverman et al. 1981a). A product containing nematodes for outdoor applications to turf has shown to be effective in reducing flea populations in soil (Hendersen et al. 1995).

Again for possible insecticidal treatment several products and ways of applications are available. The most common products used outdoors are sprays, dusts, or granules (Scheidt 1988) (for further information see again Kwochka 1987).

# Resistance

**8**

Resistance to insecticides is defined by the World Health Organization (WHO) as 'development of an ability in a strain of some organism to tolerate doses of a toxicant that would prove lethal to a majority of individuals in a normal population of the same species' (WHO, cited in Ferrari 1996). Ferrari (1996) states that 'resistance has a genetic composition of a population as a direct result of the selective effects of a toxicant'. Bossard et al. (1998) proclaim 'a response of an organism or a population to a toxicant that enables the organism or population to withstand future toxicant exposures better, because gene amplification which may confer resistance does not require selection (Devonshire and Field 1991) and other individual responses to sublethal exposures are included' to be a better definition of resistance. They suggest much of the insecticide resistance ascribed to cat fleas possibly to be simply a variation in flea susceptibility.

Resistance in cat fleas was first reported in 1952 in dogs in the southeastern United States. Thereafter cat fleas from many areas showed resistance to chlordane, dieldrin and hexachlorocyclohexane (HCH) (Brown and Pal 1971) and other insecticides (WHO 1992). Besides these first reports, numerous insecticide resistances and different resistance ratios have been published (see Table 9). But at times, the ratios are less than the expected variability within cat flea strain (Moyses 1995), indicating that adult fleas may not be resistant but inherently tolerant of e.g. certain pyrethroid formulations (Bossard et al. 1998). Whether resistance ratios are caused by strain differences or bioassay conditions is often difficult to distinguish in a number of resistance reports, because field and reference strain have not been assayed simultaneously (Bossard et al. 1998; Dryden and Broce 2000; Dryden et al. 2000). Of the species of fleas tested, *C. felis* is resistant to the greatest number of different categories of insecticides (WHO 1992).

In general, resistance ratios in fleas are low compared with other insecticide-resistant arthropods (such as horn flies against fenvalerate (92,000-fold) (Sheppard and Joyce 1992), diamondback moths resisting *Bacillus thuringiensis* (6,800-fold) (Tabashnik et al. 1993), and the tick *Boophilus decoloratus* to fenvalerate (4,744-fold) (Coetzee et al. 1987)). The low resistance ratios in fleas, coupled with high strain variability, increase the difficulty of detecting resistance (Moyses 1995).

Whether the characteristics of cat flea resistance are caused by biological, operational, or assay characteristics is unknown (Bossard et al. 1998).

Table 9. Resistance ratios as reported in the literature for the cat flea (*C. felis*)

| Insecticide | Resistance ratio | Reference |
|---|---|---|
| Bendiocarb | 28 | El-Gazzar et al. 1986 |
| | 7.7 | El-Gazzar et al. 1988 |
| Carbaryl | 12 | El-Gazzar et al. 1988 |
| | 20 | El-Gazzar et al. 1986 |
| Chlorfenvinphos | 5.6 | El-Gazzar et al. 1986 |
| | 1.3 | El-Gazzar et al. 1988 |
| Chlorpyrifos | 10 | El-Gazzar et al. 1986 |
| | 4.7 | Moyses 1995 |
| | 1.6 | El-Gazzar et al. 1988 |
| Cyfluthrin | 6.8 | Lemke et al. 1989 |
| Cypermethrin | 5.2 | Lemke et al. 1989 |
| Diazinon | 15 | Moyses 1995 |
| | 9.4 | El-Gazzar et al. 1986 |
| | 1.3 | El-Gazzar et al. 1988 |
| D-phenotrin | 1.2 | Lemke et al. 1989 |
| Fenitrothion | 5.8 and 41* | Moyses 1995 |
| Fenthion | 2.0 and 10 | Moyses 1995 |
| Fenvalerate | 2.8 | Lemke et al. 1989 |
| Fluvalinate | 4.2 | Lemke et al. 1989 |
| Isofenphos | 3.5 | El-Gazzar et al. 1986 |
| | 1.1 | El-Gazzar et al. 1988 |
| Malathion | 690* | Moyses and Buchy 1996 |
| | 190 | Moyses 1995 |
| | 77 | Kobayashi et al. 1994 |
| | 25 | El-Gazzar et al. 1986 |
| | 12* | Collart and Hink 1986 |
| | 1.7 | Moyses 1995 |
| | 1.6 | El-Gazzar et al. 1988 |
| Malathion + DEF | 12* | Moyses 1995 |
| Permethrin | 12 | Moyses 1995 |
| | 1.6 | Lemke et al. 1989 |
| Propetamphos | 7.2 | El-Gazzar et al. 1986 |
| | 1.1 | El-Gazzar et al. 1988 |
| Profenofos | 1.4 | Moyses 1995 |
| Propoxur | 4.4 | El-Gazzar et al. 1986 |
| | 2.7 | El-Gazzar et al. 1988 |
| Resmethrin | 2.2 | Lemke et al. 1989 |
| Tralomethrin | 1.5 | Lemke et al. 1989 |

* laboratory selected strain

Examining the results of resistance bioassays, a number of factors possibly contributing to variability should be considered: abiotic (temperature, humidity, lighting (Hinkle et al. 1989; Rust 1995), type of solvent (Rust 1993), volume of solvent (Moyses 1997), substrate, procedure); biotic (density of insects (Shepard 1960), colonization, sex, age and developmental stage). For further information see Bossard et al. (1998).

For some susceptibility differences in cat flea resistance, a genetic basis is suggested, relying on the results of laboratory experiments (Rust 1993). In resistant cat fleas target site insensitivity of acetylcholinesterase and organophosphate detoxification by glutathione transferase conjugation occur (Hinkle et al. 1995b). Detoxification mechanisms expected to exist in cat fleas also include detoxification of D-limonene via mixed function oxidases (MFO's) (Collart and Hink 1986) and cytochrome p450 (mixed function oxidase) detoxification (Bossard 1997). No correlation could be found between esterase levels and mortalities in nine cat flea strains when exposed to a number of insecticides (Bossard 1997). Also no resistance has been reported caused by modifications in behavior, morphology, or the excretion or sequestration of insecticides in cat fleas (Bossard et al. 1998).

Factors which generally can affect the development of resistance in insects include isolation, mobility, fortuitous survival, polyphagy, and refugia (Georghiou and Taylor 1986). Concrete for (cat) fleas interhost movement, mortality by abiotic factors such as temperature and humidity, host resistance in form of intense grooming, movements and home ranges of hosts, host variety and habitat heterogeneity will influence and affect resistance development in a positive or negative manner. (For further information see Bossard et al. 1998, Dryden and Broce 2000.) Mortality from noninsecticidal sources should slow resistance development e.g. by removing resistance genes (Rosenheim et al. 1996). The establishment of refugia by a wide variety of hosts e.g. might suppress resistance by providing susceptible fleas (Bossard et al. 1998). Exophilus resistance could result when cat fleas move to difficult-to-treat areas in response to darkness or humidity (Bossard et al. 1998).

A possible way of resistance production might be the ingestion of flea feces by larval fleas, if these were contaminated with insecticides (Bossard et al. 1998).

Resistance development is further dependent on operational factors, including the insecticidal mode of action, cross-resistance, selection intensity, and life stage selected. (E.g. in studies of El-Gazzar et al. (1988) in a resistant Florida strain, larvae have been shown to be more resistant towards some of the tested insecticides than adults and to be more susceptible towards others.) Inadequate insecticide applications, insecticides with prolonged residual activity, and sustained release devices may kill less than 100% of the pest insects and contribute to the rapid evolution of resistance (Sheppard et al. 1989; Rust 1995). With the introduction of systemical and topical formulations (of fipronil, imidacloprid, insect growth regulators and developmental inhibitors such as lufenuron (Hinkle et al. 1995a, 1997)), flea populations may be exposed chronically to sublethal doses, increasing resistance, especially when prophylactically administered (Bossard et al. 1998).

Insecticide tolerance and resistance are often blamed for failures to control cat fleas (Fox et al. 1968; Kerr 1977; Davidson 1992), but other reasons such as reinfestation from refugia on domestic and wild animals or in off-host environments (Bennett and Lund 1977; Byron 1987) should also be considered as factors contributing to failure of control (Bossard et al. 1998). Besides biological factors of the flea's life cycle management factors may also contribute to insecticide resistance. Insecticide applications that are not adjusted to variation in humidity and temperature or cultural conditions such as substrates or carpet types may cause control failures (Koehler et al. 1986; Dryden and Reid 1996). Poor penetration of carpeting by insecticides and the subsequent reemergence of adults (Dryden and Rust 1994) as well as failure to treat larval breeding habitats (Rust and Dryden 1997) have to be kept in mind. The tolerance of cat flea cocoons placed in carpet to a variety of insecticides (Rust and Reierson 1989), for example is not due to any protection afforded by the cocoon, but to the lack of penetration of the carpet canopy by the insecticide according to Dryden and Rust (1994). Dryden and Reid (1993) reported that cocoons are not a barrier to insecticides. In trials of Koehler et al. (1986) registered insecticides applied on nylon carpet provided significant mortality for only 1-7 days, possibly the result of using a resistant flea strain. In the context of these studies carpet appears to be a surface on which it is difficult to control resistant cat fleas (Koehler et al. 1986). At least three factors may adversely affect the availability of insecticide residues on carpet (Koehler et al. 1986): (1) Carpet has a greater surface area per $cm^2$ compared with most commonly treated surfaces resulting in less active ingredient per unit area. (2) Carpet density prevents complete coverage of the treated surface, and third, since nylon carpet is a synthetic organic material, synthetic organic pesticides may move into the fiber matrix and not be available to kill the target pests. The net effect of high carpet surface area per $cm^2$, carpet density preventing complete coverage, and the possible binding of insecticides in the fiber results in poor residual control of insecticide-resistant fleas (Koehler et al. 1986).

In general to determine whether cat fleas are truly resistant to insecticides baseline studies of population susceptibility before and after toxicant exposure are required, as well as unexposed populations monitored simultaneously (Bossard et al. 1998; Dryden and Broce 2000; Dryden et al. 2000).

Pest management strategies that reduce the likelihood of resistance need to be developed (Bossard et al. 1998). Flea control measures should be designed to minimize the risk of resistance developing with incorrect dosing (Dryden 1996). When adequately detailed data on local epidemiologic patterns become available, it should be possible to replace routine dosing with strategic treatment programs (Jacobs et al. 1997b).

Jacobs et al. (1997b) suggest that it is also possible that the risk of resistance could be further reduced by programs using compounds with different modes of action as is advocated for the control of anthelmintic resistance in sheep (Barnes et al. 1995).

# Imidacloprid

## History of Imidacloprid

In the 1960s the naturally occurring insecticide nicotine was simplified by Yamamoto (1965) to 3-pyridylmethylamines, but none of these were of practical value. At the IUPAC Conference in Zurich in 1978 Shell Chemical Co. (Soloway et al. 1978) introduced a new class of chemicals named nitromethylene insecticides represented by its most active member nithiazine which was never commercialized for broad agricultural use (Kollmeyer et al. 1999). In recent times a knock-down fly product against *M. domestica* containing nithiazine has been available. In 1985 finally the chemists of Nihon Tokushu Noyaku Seizo K.K. in Japan synthesized among other compounds the nitroguanidine NTN 33893, known today as imidacloprid (Fig. 13 and Fig. 14) (Elbert et al. 1998; Nauen et al. 2001). The addition of a heterocyclic substituent, the 6-chloro-3-pyridyl-methyl moiety, to the nitromethylene substituted heterocycles (NMHs) increased the insecticidal activity remarkably (Kagabu 1996; Yamamoto et al. 1998) and further optimization finally resulted in the invention of imidacloprid from Nihon Bayer Agrochem. K.K. (Shiokawa et al. 1986). Thus imidacloprid was the first commercialized member of a new class of insecticides, called chloronicotinyl or neonicotinoides (Nauen et al. 2001; Yamamoto et al. 1998).

The inactivity of the benzyl substituted nitromethylenes determined by Soloway et al. (1979) was a first hint that the effectiveness of imidacloprid depends among others on the configuration of the pyridyl (Leicht 1996) (see Fig. 13 and Fig. 14). Surprisingly, a substituted chloropyridyl compound was also discovered in the skin of an arrow-poison frog (*Epipedobates tricolor*) from Ecuador and was named epibatidine (Fig. 14). It was characterized as a pain-killer 200-times more effective than morphine (Spande et al. 1992).

Due to Bayer's success with imidacloprid several other companies such as Takeda, Nippon Soda, Agro Kanesho, Mitsui Toatsu and Novartis introduced their own neonicotinoid insecticides. These chemicals only differ regarding the heterocyclic group or concerning the structure of the pharmacophore and their insecticidal properties are comparable to those of imidacloprid (Nauen et al. 2001).

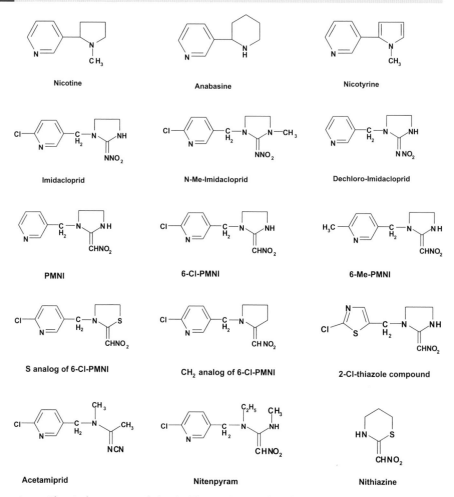

Fig. 13. Chemical structures of nicotinoids, neonicotinoids and their related compounds

## Chemical Properties of Imidacloprid

Imidacloprid belongs to the group of neonicotinoid insecticides. Its physicochemical properties are listed in Table 10. For further information on physical and chemical properties see Diehr et al. (1991).

One intermediate for the synthesis of imidacloprid is 2-chloro-5-chloromethylpyridine (CCMP). Out of this compound as well as ethylenediamine and nitroguanidine the active ingredient named imidacloprid is synthesized (Diehr et al. 1991).

Generally, subtle differences in chemical structure could provide differences in the following characteristics: (1) photostability, (2) persistence/residual activity, (3) meta-

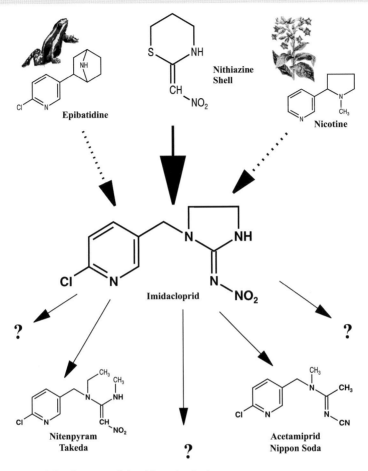

Fig. 14. Chemical development of the chloronicotinyles

bolism, (4) partition coefficient, and (5) target pests and use pattern. The nitro-methylenes reported by Soloway et al. (1979) showed rapid hydrolysis and were photolabile (Mullins 1993). Compared to nithiazine imidacloprid exhibits a good level of photostability that is appropriate for a variety of uses including outdoor agricultural use (Mullins 1993). Testing the compound against a number of insect pests, e.g. the green peach aphid (*Myzus persicae*), the cotton whitefly (*Bemisia tabaci*), the green rice leafhopper (*Nephotettix cincticeps*) or the Colorado potato beetle (*Lepinotarsa decemlineata*) imidacloprid was highly active and possessed a positive correlated temperature coefficient. This means that with rising temperature the compound showed increasing biological activity (Elbert et al. 1991). The most important properties can be summarized as follows (here with special emphasis on agricultural use) (Mullins 1993):

Table 10. Physicochemical properties of imidacloprid

| Empirical formula | $C_9H_{10}ClN_5O_2$ |
|---|---|
| Molecular weight | 255.7 |
| Physical appearance | colorless crystals |
| Melting point | 144°C |
| Solubility at 20°C (g a.i./1000 ml) | water: 0.61 |
| | n-hexane: <0.1 |
| | 2-propanol: 1-2 |
| Vapor pressure | $4 \times 10^{-7}$ mPa at 20°C |

a.i.: active ingredient

- broad spectrum activity/broad crop applicability
- favorable characteristics for applicator/consumer/environment
- new mode of action for insecticide resistance management
- new tool for integrated pest management

As can be seen from the structural formula there is no chemical relation between nicotine and imidacloprid (Fig. 14). Both nicotinoids and the so-called neonicotinoids, to which among others imidacloprid belongs, are characterized by the presence of a pyridyl moiety in their structure (Yamamoto et al. 1998). Nicotinoids further possess an amino nitrogen which is ionized, while in neonicotinoids the corresponding nitrogen atom is not ionized but bears a partial positive charge. The partial positive charge on a nitrogen atom of neonicotinoids which is conferred by the neighboring electron-withdrawing group such as nitromethylene, nitroimine, or cyanoimine (Yamamoto et al. 1995) can distinguish the insect receptor from the vertebrate nicotinic acetylcholine receptor (nAChR) which is generally the side of action of imidacloprid (further information see chapter 'Mode of Action and Direct Insecticidal Activity'). For penetration into the insect central nervous system, hydrophobicity seems to play an important role (Yamamoto et al. 1998). In contrast to nicotine, imidacloprid is non-ionized and more hydrophobic, which seems to be a factor in increasing the translocation into the insect´s central nervous system (Yamamoto et al. 1998). The introduction of the chlorine atom at the 6-position on the 3-pyridyl group also increases the hydrophobicity (Yamamoto et al. 1998), and the binding affinity (Tomizawa and Yamamoto 1993), in the course of which the former effect seems to be more important (Yamamoto et al. 1998). The significance of the chlorine atom or methyl group is based on the apparent increase of the translocation into the insect´s central nervous system by increasing hydrophobicity (Yamamoto et al. 1998), but the possibility of metabolic masking was also considered for them (Tomizawa and Yamamoto 1993). Among neonicotinoids, nitromethylene type compounds, though far higher in binding affinity, were less hydrophobic than the corresponding nitroimine type (Yamamoto et al. 1998). To be insecticidal, nitroimino type compounds must have higher hydrophobicity because of their lower binding affinity, and nitromethylene compounds, though potent in binding affinity, must have reasonable hydrophobicity

(Yamamoto et al. 1998). Last but not least not only a better penetration into the insect´s central nervous system, but also an improved penetration through the cuticle may be possible because of the increased hydrophobicity (Yamamoto et al. 1998).

Effects of different substituents and modifications on the individual structures of NMHs on the structure-activity relationships (SAR) of nicotinoids and imidacloprid analogues can be found in detail in Tomizawa and Yamamoto (1993).

Summarizing the essential differences between nicotinoids and neonicotinoids, the former possess the essential feature of a basic nitrogen which becomes positive by protonation in the insect body, thus causing the molecules to resemble acetylcholine. The latter, imidacloprid, PMNI and related compounds share the same binding site, the same essential moiety and similar structure-activity relationships as nicotinoids, but their nitrogen possesses a partial positive charge due to the electron-withdrawing effect of the substituents on 6-position therefore proposing a different name for distinguishing them from nicotinoids by the term neonicotinoids (Tomizawa and Yamamoto 1993). The cationic ammonium head of nicotinoids is required for interaction at the anionic site on the receptor, but restricts the penetration of the compounds through the cuticle and into the nerve because of the ionic form, thus explaining the rather low insecticidal activity of classical nicotinoids (Tomizawa and Yamamoto 1993). In neonicotinoids the unshared electron pair on the nitrogen atom corresponding to amino nitrogen atom of nicotinoids is delocalized by the strong electron-withdrawing effect of the neighboring substituents. In other words, the concerned nitrogen atom becomes only partially positive (Yamamoto et al. 1995).

# Biological Profile, Mode of Action and Direct Insecticidal Activity

Certain types of neuronal receptors respectively ion channels have been demonstrated to be major target sites of insecticides (Narahashi 1996). Imidacloprid acts agonistically on nAChRs. However, nAChRs represent a new target site considering commercial importance, if one looks into the market importance of insecticides of the categories organophosphates, carbamates and pyrethroids (Naumann 1994) (see Fig. 15).

## Biological Profile

Imidacloprid has proven to possess strong insecticidal activity. Insecticidal activity in general is an overall expression of intrinsic activity at the site of action, activation and detoxication, excretion, penetration, etc. (Tomizawa and Yamamoto 1993). It further is characterized as an integrated effect comprising intrinsic activity at the site of action, in vivo behavior like excretion or penetration, and stability on the host plants in the open as well as in the soil (Kagabu and Medej 1995). The insecticidal activity of compounds should be governed by various factors (Nishimura et al. 1994): penetration through the cuticle, transport within the insect body, metabolic inactivation, binding to the target sites, intrinsic activity at the target sites, and others.

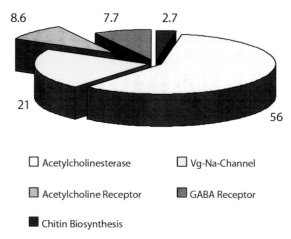

8.6        7.7     2.7

21

56

☐ Acetylcholinesterase        ☐ Vg-Na-Channel

▨ Acetylcholine Receptor      ▨ GABA Receptor

■ Chitin Biosynthesis

**Fig. 15.** Mode of action of the major insecticides in relation to their sales volume (1997) in the agrochemical market (excluding fumigants and those compounds with unknown mode of action) (vg: voltage-gated)

The biological profile of imidacloprid was defined on the basis of the results of exhaustive laboratory experiments and greenhouse trials (Elbert et al. 1991). The spectrum of activity primarily includes sucking insects (aphids, whiteflies, leaf- and planthoppers, thrips, plant bugs, and scales), and also many Coleopteran species (Colorado potato beetle, white grubs, leaf beetles, etc.) and selected species of Lepidoptera (Elbert et al. 1991). Activity has also been demonstrated for selected species of Diptera, for ants (Hymenoptera), termites (Isoptera), cockroaches, grasshoppers and crickets (Orthoptera) (Mullins 1993). Due to its favorable mammalian safety characteristics it has been developed for veterinary use (Werner et al. 1995a), thus being the first compound to be developed commercially for animal use from the new chemical class of chloronicotinyl nitroguanidine insecticides (Griffin et al. 1997).

## Mode of Action and Direct Insecticidal Activity

To illustrate the mode of action of the compound imidacloprid some general information on neurotransmission is presented first:

The cells of the nervous tissue (neurons) consist of a cell body with a nucleus, an axon and dendrites. The ramifying axons and dendrites come into contact with other neurons or muscle fibers which is especially the case in the periphery, and form so-called synapses. These synapses generally possess a chemical principal in which information from one neuron to another one or to a muscle fiber is transmitted via a chemical compound, the so-called transmitter. These transmitters are stored within the presynaptic part of the neuron in 'pools'. An incoming action potential releases the transmitters out of their presynaptic pools into the synaptic cleft. The binding of the transmitters to receptors of the postsynaptic membrane causes a specific change in the conductibility of the membrane for special ions, resulting in a change of the postsynaptic potential based on altered permeability of the neutral membrane. For simplicity, complexes which represent the molecular targets within the nervous system in form of voltage- and ligand-activated ion channels and the associated ligand recognition sites are termed 'receptors'. These receptors are encoded by multigene superfamilies (Benson 1989). The permeability changes result in the generation of a new nerve impulse in the postsynaptic cell. At the cholinergic synapses (see below) the binding process is terminated by the enzyme acetylcholinesterase, which inactivates the neurotransmitter acetylcholine (ACh) by hydrolyzing it into acetate and choline, the latter being resorbed by the presynaptic neuron and used in the synthesis of new neurotransmitter by the enzyme choline acetyl transferase.

At the motoric terminal plate, the connection between neuron and muscle fiber, the postsynaptic action potential induced by depolarization causes a release of $Ca^{2+}$ from vesicles (in the longitudinal tubuli) thus resulting in contraction of the myosin/actin complex of the sarcomers.

Within the insect nervous system a number of transmitters can be differentiated of which acetylcholine (ACh), γ-aminobutyric acid (GABA), glutamate, serotonin and octopamin are the best examined compounds (Abbink 1991). ACh is the most important excitatory neurotransmitter of the central nervous system (CNS) of insects (Methfessel 1992). In AChRs two different classes have been identified throughout the animal phyla, including insects: nicotinic (nAChR) and muscarinic (mAChR) ones (Sattelle 1980; Venter et al. 1988). Respectively for the insect nervous system some authors even describe at least three populations of binding sites which display solely nicotinic, solely muscarinic, and mixed nicotinic-muscarinic pharmacology (Donnellan et al. 1979; Sattelle 1980; Eldefrawi and Eldefrawi 1983). Nicotinic AChRs consist of the ACh agonist site and an ion channel complex (Devillers-Thiéry et al. 1993; Galzi and Changeux 1994). They are present in much greater concentration than muscarinic receptors. Insect nervous tissue appears to be the richest source in the animal kingdom of neuronal nAChRs (Breer and Sattelle 1987). In insects, the nAChR was thought to be exclusively located in ganglia (Breer and Sattelle 1987), but findings of stationary immobility and intense continuous tetanus-like trembling in fleas after being exposed to imidacloprid suggest that there may be additional blocking processes, generally at the connections between nerves and muscles, the so-called motoric endplates (Mehlhorn et al. 1999). Mammalian neuronal nAChRs are found in both the central and peripheral nervous systems (CNS and PNS). It is known that drugs interact with nAChR in three ways (Tomizawa et al. 1995a):

1. Agonistic action, that is activation of the ion channel by drug binding to the ACh recognition site.
2. Competitive blockade action, that is inhibition of the activation of the ion channel by drug binding to the ACh recognition site.
3. Noncompetitive blockade action, that is inhibition of the ion channel by drug binding to the allosteric site located in the receptor's ion channel.

The nAChR is a homo- or heteropentameric membrane protein, the subunits of which form a ligand-gated cation channel. The nAChR protein comprises a mixture of $\alpha$ and non-$\alpha$ subunits. Since there are multiple types of both subunits, considerable molecular heterogeneity can exist at the level of the whole protein (Lind et al. 1998). The vertebrate nAChR has been characterized as a pentameric ligand-triggered cationic ion channel with a central pore (Stroud and Finer-Moore 1985). According to Matsuda et al. (1999) the nAChR-heteropentamers are composed of two $\alpha$ and three non-$\alpha$ subunits surrounding a central cation-selective ion channel that opens transiently in response to the binding of ACh or nicotinic agonists. In insects, a number of subunit cDNAs have been cloned (reviewed by Gundelfinger 1992). Six nAChR subunits from insect species have been studied in detail (Matsuda et al. 1999).

In vertebrates, it has been found that while the subunit composition of nAChRs at adult neuromuscular junctions is invariable, there is considerable diversity of neuronal nAChRs (McGehee and Role 1995). While all nAChRs interact with ACh, physiological and pharmacological studies have revealed the existence of a number of receptor subtypes (McGehee and Role 1995). In mammals they show a considerable diversity due to the assembly of different receptor subunits into pentameric complexes (reviewed by Galzi et al. 1991; Lukas and Bencherif 1992; Role 1992; Sargent 1993). The distinct subtypes of mammalian nAChRs again show distinct physiological properties as well as different sensitivities to agonists and antagonists (Cachelin and Jaggi 1991; Gross et al. 1991; Luetje and Patrick 1991; Luetje et al. 1990; Mulle et al. 1991; Papke et al. 1989). A similar complex picture of insect-type nAChRs is emerging (reviewed by Gundelfinger 1992; Nauen et al. 2001).

Differences among receptors belonging to the same family but occurring in different phyla, if not species, seem to be sufficient to provide an opportunity for exploitation by pesticide chemists (Benson 1989). According to molecular biological studies nAChRs of insects and vertebrates show a close structural relation, but are different from one another in detail (Breer and Sattelle 1987; Gundelfinger 1992).

Based on data about the complexity of nAChR of *Myzus persicae* (Huang et al. 1999; Sgard et al. 1998) and *Locusta migratoria* (Hermsen et al. 1998) it was suggested that the insect nAChRs may indeed be as complex as their vertebrate counterparts (Huang et al. 1999). Phylogenetic comparisons indicate that the insect nAChRs are divided into two major groups, $\alpha$ and non-$\alpha$ (Huang et al. 1999), and that insect $\alpha$ subunits appear to have evolved in parallel to the vertebrate neuronal nAChRs (Huang et al. 1999; Le Novere and Changeux 1995; Ortells and Lunt 1995; Tsunoyama and Gojobori 1998).

Nicotinic AChRs have been exploited as an insecticidal target with the insecticides nicotine and cartap (a nereistoxin analogue) and more successfully with the introduction of a chloronicotinyl compound, the nicotinic agonist imidacloprid (Corbett et al. 1984; Elbert et al. 1990).

From contact and oral administration, the mode of action of nitromethylene compounds is mediated via the nervous system of insects (Schroeder and Flattum 1984).

For better examination of the identified target site of nitromethylenes, the compound ($\alpha$-bungarotoxin ($\alpha$-BGT) is used. The [$^3$H]$\alpha$-BGT binding site as ACh recognition site on the insect AChR was described by Sattelle et al. (1983) and Lummis and Sattelle

(1985). As a specific ligand of the nAChR α-BGT is displaced from its target site in ganglia of cockroaches (Bai et al. 1991) and housefly head homogenates (Abbink 1991) by imidacloprid. In the nAChRs from *Torpedo* electric organ, a modified muscle, and from vertebrate skeletal muscle α-BGT also has been generally understood to bind specifically to the ACh recognition site as a competitive blocker (Albuquerque et al. 1979).

Schroeder and Flattum (1984) were the first who identified the nAChR as molecular target of nitromethylenes (by using nithiazine). They were the first to determine the site of action of NMHs using the American cockroach, *Periplaneta americana*, explicitly the cockroach´s sixth abdominal ganglion with its cercal nerve-giant fiber synapses. Neurophysiological experiments with NMHs caused a biphasic effect, characterized by an initial increase in the frequency of spontaneous giant fiber discharges thus representing a stimulation of the nervous system, followed by the development of a complete block to nerve impulse propagation, in short a conduction block (Schroeder and Flattum 1984). The specific site of action of the NMHs was presumed to be postsynaptic, with probable involvement of an agonistic effect on postsynaptic AChRs. In the trials the effects of NMHs on the cercal nerve-giant fiber synapses of the American cockroach did not prove to be an inhibition of acetylcholinesterase (I) or a disruption of the mechanism of neurotransmitter release (II), but hypothesized to be a direct agonist effect on postsynaptic AChRs (III) (Schroeder and Flattum 1984). Schroeder and Flattum (1984) proposed that the NMHs act as neurotransmitter mimics to activate and then block the postsynaptic AChRs. Their studies showed that NMHs have both excitatory and depressant effects on nerve transmission at the cholinergic synapses between the afferent cercal sensory nerves and the giant interneurons of the cockroach. Further studies in cockroaches concentrating on an identified motor neuron, the fast coxal depressor motorneuron $D_f$, verified the agonist action of imidacloprid in form of an interaction with insect neuronal nicotinic receptors (Bai et al. 1991). Numerous authors using different insect species observed the effect of imidacloprid on the nAChR (honeybee heads (Tomizawa et al. 1995b), houseflies (Abbink 1991; Liu and Casida 1993), locust thoracic ganglion (Zwart et al. 1994), fruit fly, honey-bee, milkweed bug, German cockroach, American cockroach and cricket (Liu and Casida 1993).

Relying on the results of various studies imidacloprid is known to interact with the nAChR (see authors above and Tomizawa et al. 1995a, 1995b). For the insecticidal activity Nauen et al. (1999) found the agonistic action to be a prerequisite and to be essential while antagonistic compounds were mostly non-active. But recent studies hypothesized both agonistic and antagonistic effects on nAChR channels by imidacloprid (Nagata et al. 1998).

After determining the site of action of NMHs respectively imidacloprid as the postsynaptic nAChR, the general mode of action can be described as follows: Imidacloprid acts by interfering with the transmission of nerve impulses in insects (Griffin et al. 1997). The binding to the nAChR in the postsynaptic region of the insect nerve leads to an opening of a sodium ion channel (Mullins 1993). In contrast to acetylcholine, imidacloprid is only slowly degraded at the postsynaptic nAChR, producing a prolonged action that disrupts normal nervous system operation. This substantial disorder within the nervous system leads to consequent lethal action in most cases (Mullins 1993). Particularly in cat fleas the continuous blockage of the nAChR manifests in light- and electron-microscopical changes with the compound presumably taken up via the smooth, nonsclerotized intersegmental membranes. Muscles and nerves are irreversibly destroyed, with the head and thorax ganglia and the striated muscles being affected first. Degeneration in form of destruction of the mitochondria, damage to the nerve cells, and disintegration of the muscles occurs. In detail lightening of the zone of neurons in light-microscopy and disruption of the cytoplasm, vacuolization of most mitochondria, swelling of the perinuclear space, and degradation of the nuclear contents in electron-microscopy could be observed (Mehlhorn et al. 1999) (Fig. 38, Fig. 39).

Yamamoto (1996) describes the neonicotinoid mechanism as agonist action on the nAChR at the synapses in the insect central nervous system with a stimulation of the postsynaptic membrane first and then a paralyzing effect on nerve conduction. Sone et al. (1994) observed a biphasic effect of imidacloprid: initial stimulation in form of strong repetitive firing as agonist action in the American cockroach finally followed by a cessation of the nerve conduction, corresponding with Bai et al. (1991) who also suggested the conduction block to be caused by the binding of imidacloprid to the cholinergic receptors in the nerve membrane.

In more detail, an inward current elicited by NMHs resembling inward currents induced by ACh was observed by Cheung et al. (1992). A slow depolarization is caused by imidacloprid (here in the cockroach motorneuron) (Bai et al. 1991). Zwart et al. (1994) observed nAChR-mediated inward currents induced by nitromethylenes with a final blocking of these inward currents in continued presence of the NMHs. Agonist and blocking actions of the NMHs appear to be related and thus Zwart et al. (1994) concluded that NMHs act as agonists on the locust nAChR and that the blocking effect is due to desensitization, a common effect of agonists on nAChRs. In further studies single-channel currents of multi-conductance states using rat phaeochromocytoma (PC12) cells were generated by imidacloprid (Nagata et al. 1996). Both, ACh and imidacloprid, induced single-channel currents of main and subconductance states, in the course of which ACh predominantly generated the main conductance currents and imidacloprid the subconductance currents (Nagata et al. 1996, 1997). When applied together, both conductance states were generated, possibly indicating that the two compounds act on the AChR independently without interacting with each other (Nagata et al. 1997). It was shown that imidacloprid opens the main conductance and subconductance channels (in PC12 cells) which are the same as those opened by ACh in addition to at least two other conductance levels (Nagata et al. 1998). Because of the much smaller maximum current induced by imidacloprid in the mammalian PC12 cells compared to other compounds (here carbachol), imidacloprid is considered to be a partial agonist (Nagata et al. 1998). It is thought that the molecule binding to the agonist binding site may cause changes in confirmation that result in an open channel with reduced conductance as well as an open channel with full conductance (Nagata et al. 1998). In these trials using mammalian neuronal AChRs, imidacloprid had both multiple agonist and antagonist effects on the neuronal nAChR-channels. Hypothesizing two binding sites with different affinities for imidacloprid and possibly both agonistic and antagonistic effects on the nAChR channels (Nagata et al. 1998) the original classification has slightly changed and further investigation has to follow.

High affinity binding of radiolabeled imidacloprid was observed in housefly head membranes in form of a rapid biphasic association and dissociation consistent with a two-stage sequential reaction in each case (Liu and Casida 1993). Lind et al. (1998) have observed the existence of two imidacloprid binding sites with high and low affinity in two hemipteran species, the peach potato aphid *M. persicae* and the green leafhopper *Nephotettix cincticeps*, whereas in other insects, non-hemipteran species such as *Drosophila*, houseflies and also *C. felis* only a single site was observed. This may indicate that in these insects only a single class of binding site is present and/or that there is no cooperativity between binding sites (Lind et al. 1998). Non-hemipteran insects either do not possess multiple imidacloprid binding sites or the heterogeneity of nAChRs is not revealed in these insects by imidacloprid (Lind et al. 1998). The presence of very high-affinity binding sites only in hemipteran insects may explain the particular usefulness of the compound in controlling sucking pests including aphids and plant hoppers (Lind et al. 1998). Nevertheless Liu and Casida (1993) found specific binding in each insect they examined (including among others fruit fly *Drosophila melanogaster*, honey-bee *Apis mellifera*, milkweed bug *Oncopeltus fasciatus*, German cockroach *Blatella germanica*, housefly *Musca domestica*, American cockroach *Periplaneta americana*, cricket *Teleogryllus commodus*) with particular high activity for housefly head as well as cricket and American cockroach brain.

Even though Liu and Casida (1993) could not detect an imidacloprid binding site in vertebrate brain or electric organ of the electric eel in their trials, Zwart et al. (1994) suggest the same mechanisms of action of NMHs on mammalian endplate-type as on locust neuron nAChRs. But much higher concentrations of NMHs are required to activate and desensitize endplate-type nAChRs (Zwart et al. 1994). Poor interaction of nitromethylenes or chloronicotinyls with mammalian nAChRs and thus low mammalian toxicity was shown in a number of studies using rat muscle (Methfessel 1992), various vertebrate brains (human, dog, mouse, chicken) and electric eel electric organ (Liu and Casida 1993), mouse cells (N1E-115 and BC3H1 neuroblastoma cells; Zwart et al. 1994) and *Torpedo* electric organ (Tomizawa et al. 1995a) just to mention some of the tested cell systems. These results are supported by toxicological studies in various vertebrate species (see chapter 'Toxicology of Imidacloprid'). Compared to nicotine, imidacloprid has also a lower mammalian toxicity (Yamamoto et al. 1995), which may be due to a lower affinity to the ACh recognition site (Tomizawa et al. 1995a), with the suggestion being based on the weaker agonist action of neonicotinoids on nAChR of *Torpedo* electric organ (Tomizawa et al. 1995a) or similar of rat brain (Yamamoto et al. 1995).

Selective toxicity can be exerted by different ligands based on various subtypes of nAChRs exhibiting significant differences in sensitivity towards nicotinyls (Yamamoto et al. 1995). Nauen et al. (1999) even confirmed the existence of functional nAChR with distinct affinity and different ion channel properties and pharmacology in isolated locust neurons, making them suggest the existence of (at least) two different and independent populations of nicotinic receptors in the insect nervous system, in accordance with conclusions of Van Den Beukel et al. (1998). AChRs possess different pharmacological properties within one species. But also between homologous receptors of different species there exists considerable differences in sensitivity towards agonists and antagonists (Leicht 1996). Not only different metabolic or physiological facts, but also the molecular target, the nAChR, causes the advantageous species selectivity (Methfessel 1992). The relation between chemical structure and physiological effect of the NMHs appears to depend on nAChR properties, which vary between the different subtypes of receptors (Zwart et al. 1994). The physicochemical properties of chloronicotinyls may not only be responsible for selective insect affinity but also for arthropod species-selective effects (Londershausen 1996).

The binding of imidacloprid to structures of the nAChR is the subject of various studies. Tomizawa et al. (1995b) reported of a far higher affinity of imidacloprid for the ACh recognition site than for the allosteric site located in the receptor's channel, whereas nicotinoids act on both the ACh recognition and the allosteric sites. Only one molecule of the compound imidacloprid seems to be necessary to cause a channel opening (Methfessel 1992). Methfessel (1992) speculated whether this was caused because the binding site of the second $\alpha$ subunit was hidden by the structure of the molecule or because a second channel opening could not be registered. In general the $\alpha$ subunit is suggested to strongly influence and contribute to the high affinity of insect nicotinic receptors for imidacloprid (Huang et al. 1999; Matsuda et al. 1999). Examining recombinant nAChR $\alpha$ subunits of *M. persicae* in detail imidacloprid targeted selectively only two of the subunits (Mp$\alpha$2 and Mp$\alpha$3), but not the one titled Mp$\alpha$1 thus suggesting that these three subunits contribute to at least two distinct receptor populations in vivo which are imidacloprid-sensitive and -insensitive (Huang et al. 1999).

The selectivity of imidacloprid and comparable compounds for certain molecular targets may become an important tool to characterize new insect-selective molecular targets: a subtype of nAChR, which is essential for insect neurofunction but which is different in pharmacology and tissue distribution from all mammalian nAChRs investigated so far (Liu and Casida 1993).

The selective action of imidacloprid especially compared with nicotinoids such as nicotine was well examined in detail concentrating on the chemical structure, state of ionization and hydrophobicity of the compounds. Matsuda et al. (1999) suggest the 2-nitroimino-imidazolidine moiety of imidacloprid and related analogs to play a role in the partial agonist action of neonicotinoids. To interact with the insect nAChR, the presence of 3-pyridylmethylamine moiety having a positive charge (full or partial) on the nitrogen atom is essential (Yamamoto et al. 1998). This essential feature of the basic nitrogen, which becomes positive by protonation in the insect body, thus making the molecule resemble ACh, is required in form of a cationic head for interaction at the anionic site on the receptor (Tomizawa and Yamamoto 1993). The full positive charge of the amino nitrogen of the imidazolidine ring seems to be essential to interact with the vertebrate nAChR. An electron-withdrawing group such as nitromethylene, nitroimine, and cyanoimine can confer a partial positive charge on a nitrogen atom of the imidazolidine ring of the neonicotinoids (Yamamoto et al. 1995). The partial positive charge of the nitrogen atom in neonicotinoids seems to be enough to interact with the insect nAChR but not with the vertebrate receptor (Yamamoto et al. 1998). In other words, the partial positive charge in neonicotinoids can distinguish the insect receptor from the vertebrate nAChR (Yamamoto et al. 1998).

The interatomic distances between an imidazolidyl nitrogen and the pyridyl nitrogen are 5.9Å, which is a suitable value to bind to the recognition site on nAChR (Kagabu 1997). It has been proposed that the nitrogen atom of the imidazolidine ring (at 1-position) and of the pyridin ring interact with the electron-rich and hydrogen-donating sites on the nAChR (Yamamoto et al. 1995; Okazawa et al. 1998). An electron-donating atom for hydrogen bonding (here e.g. the 3-pyridyl nitrogen atom) at 5.9Å apart from the fully or partially positive charged center is required for potent interaction with the nAChR of insects (Yamamoto et al. 1995). In mammals, the effect on imidacloprid on any nAChR is weak, while nicotine affects mostly peripheral nAChRs resulting in higher toxicity.

The chlorine atom at the 6-position of the 3-pyridyl moiety may further support the resulting insecticidal activity (Nishimura et al. 1994) either because of an increase of the molecule-receptor site interaction or because of a substituent-mediated retardation of enzymatic attack at the 6-position of the 3-pyridyl group as suggested by Tomizawa and Yamamoto (1993).

A correlation between penetrability and hydrophobicity is suggested (Yamamoto et al. 1998). The non-ionized and more hydrophobic status of imidacloprid compared to nicotine seems to be a factor in increasing the translocation into the insect central nervous system (Yamamoto et al. 1998). Introduction of the chlorine atom increased the hydrophobicity of the chloronicotinyls (Yamamoto et al. 1998). And this increase of hydrophobicity seems more important than the binding (to the target site of a compound) (Yamamoto et al. 1998). The ionization reduced hydrophobicity and limited the penetration of nicotinoids, resulting in less insecticidal activity in these compounds (Yamamoto et al. 1998). The significance of the chlorine atom or methyl group is thought to be based in an increase of the translocation into the insect central nervous system by increasing hydrophobicity (Yamamoto et al. 1998). For further information on structure-activity relationships see chapter 'Chemical Properties of Imidacloprid'.

Summarizing the illustrated results and suggestions as well as information from binding assays and structure-activity relationship considerations, imidacloprid seems to be the best compromise for the following reasons gathered by Tomizawa and Yamamoto (1993):

- Imidacloprid seems to contain a moiety as proposed to be essential for nicotinoids to bind with nAChR and exert insecticidal activity. (The electron-withdrawing group on 2-position would make the nitrogen on the 1-position of the imidazolidine ring partially positive which is enough to interact at the anionic site of the receptor.)
- The imidazolidine ring seems to give the highest binding affinity among similar heterocycles.
- Being inferior to $=CHNO_2$ in binding affinity, $=NNO_2$ on the 2-position of the imidazolidine ring seems metabolically and environmentally more stable and provides binding affinity sufficient for insecticidal activity.

## Indirect Insecticidal Effects

Apart from the direct insecticidal activity of imidacloprid pointed out in the previous chapter with lethal effects demonstrated in various insect species, the compound also possesses a number of sublethal side effects.

Laboratory and greenhouse studies have demonstrated that imidacloprid also affects many insect species sublethally. These effects are sometimes dosage dependent and include: (1) repellency, (2) reduction or cessation of feeding, (3) reduction or cessation of reproductive activities, (4) overall reduction of movement or activity, and (5) increased susceptibility to biological control (Mullins 1993).

Laboratory studies with some insect species have demonstrated that sublethal doses of imidacloprid increase the insects' susceptibility to certain microorganisms that naturally occur or that could be used commercially in conjunction with imidacloprid treatments (Mullins 1993). The potential of utilizing the sublethal effects in conjunction with other control technologies integrated in pest management systems is just beginning to be explored (Mullins 1993). For larvicidal effects etc. in fleas see chapter 'Larvicidal Effect of Imidacloprid and Efficacy Enhancement (with PBO)'.

Antifeedant effect and reduced reproduction in aphids in terms of decreasing total numbers of nymphs, the latter even at concentrations where no adults were killed and most aphids were unaffected in terms of feeding behavior were observed in trials with *M. nicotianae* and *M. persicae* (Devine et al. 1996). In their studies nicotine-tolerant strains of the two aphid species also showed reduced antifeeding effect and reduced fecundity. Behavioral effects have been described, mainly in aphids by Dewar and Read (1990), Knaust and Poehling (1992) and Woodford and Mann (1992). The most detailed examinations on aphids with sublethal concentrations of imidacloprid have been performed by Nauen (1995). Using a combination of biological, analytical, electrophysiological, and video-optical methods they observed substantially altered behavior of *M. persicae*, treated with systemically applied low concentrations of imidacloprid. In detail among others low concentrations of imidacloprid induced a reduction of honeydew excretion, loss in weight, restless behavior, moving from treated to untreated leaves, reduced reproduction and eventually death due to starvation. The symptom of inhibited production of larvae of *M. persicae* by imidacloprid was already described by Lowery and Sears (1986) and Gordon and McEven (1984). In contrast to the symptoms of poisoning at higher concentrations honeydew depression and weight loss were reversible symptoms in *M. persicae* once they were offered untreated leaves 24 hours after starting to feed them on treated ones (Nauen 1995). All these facts may suggest a concentration-dependent change of the mode of action for imidacloprid on *M. persicae* (Nauen 1995). The reversible antifeedant response is possibly also triggered by the inhibition of the nAChR or by the inhibition of another target, responsible for the probing of food quality. An inhibition of the cholinergic system cannot be excluded. On the other hand, perhaps the epipharyngeal chemoreceptors are blocked by imidacloprid or the aphids lose their ability to estimate food quality in general due to inhibition of the muscles which coordinate feeding or salivation (Nauen 1995). It is so far a hypothesis that the blocking of the nAChR is responsible for both the reversible antifeeding response and the distinct irreversible symptoms of a compound acting on the insect nervous system (Nauen 1995) and it still remains unclear whether the property is related to low-level intoxication, or to another still unidentified mechanism of action (Nauen et al. 1998b).

At very low concentrations imidacloprid was further observed to alter the behavior of termites (*Reticulitermes flavipes*), by inducing no phototrophic effect followed by an antifeeding effect leading to death (Leicht 1993) and causing long term disruption of social behavior making them susceptible to microbes (Boucias 1996).

Apart from aphids where antifeedant effects of imidacloprid leading to honeydew suppression and death have been observed (Nauen 1995; Devine et al. 1996), affected feeding behavior caused by the treatment of leaf discs with sublethal concentrations of imidacloprid by foliar application as well as by systemic methods was also observed in different strains of *Bemisia tabaci*, the tobacco whitefly (Nauen et al. 1998a). In contrast to *M. persicae* with weight loss and subsequent death from starvation (Nauen 1995) in *B. tabaci* a loss in weight associated with starvation was not investigated. Nevertheless the antifeedant effect in the form of reduced honeydew expression was prominent and apparently not affected by mechanisms conferring reduced susceptibility, providing some evidence that the primary mode of action of imidacloprid on nerve transmission is not coupled with general effects of feeding behavior (Nauen et al. 1998a). In their trials Nauen et al. (1998a) observed that the larger the amount of imidacloprid ingested, the stronger the effect on feeding behavior on *B. tabaci*. They strongly suggested that low concentrations of imidacloprid, especially when distributed in host plant tissues by systemic mode of application, have a marked effect as an antifeedant on *B. tabaci* (Nauen et al. 1998a). Concentrations of imidacloprid for a 50% reduction of honeydew excretion were 16- to 852-times lower than concentrations causing 50% mortality depending on strain and application method (Nauen et al. 1998a). The tobacco aphid *M. nicotianae* not only showed less susceptibility to imidacloprid in general, but also less susceptibility to the antifeedant potential of imidacloprid in contact bioassays (Nauen et al. 1998b).

Despite all possible positive effects of sublethal concentrations of imidacloprid it is not possible to reduce the recommended field rates, because long-term field studies, resistance management strategies and the different dose-response characteristics of pests in the spectrum covered by imidacloprid have to be taken into consideration (Nauen 1995).

Studies were also undertaken to assess possible antifeeding properties of imidacloprid on fleas (Franc and Cadiergues 1998). Antifeeding action here defined as the ability of the agent to prevent the fleas from actually biting the host (Franc and Cadiergues 1998) in order to avoid the injection of antigenic saliva into the host's blood and thus prevent a rise of FAD could not be observed with the topical application of imidacloprid or fipronil. Both did not prevent fleas from biting and feeding during the first hour after infestation (prior to being killed although these products have a long term effect when observed 24 hours after infestation) (Franc and Cadiergues 1998). The topical application of dichlorvos/fenitrothion and permethrin led to a decrease in the number of engorged fleas of >80% for three and seven days post treatment, respectively (Franc and Cadiergues 1998). In contrast, recent in vitro as well as in vivo studies published by Rust et al. (2001) demonstrated the advantage of antifeeding properties of imidacloprid in cat fleas.

# Resistance of Sucking Pests Against Imidacloprid

By definition resistance is a genetic alteration of a population as response on selection by an insecticide and can impair the control in the field at recommended concentrations (Sawicki 1987). In general three genetic mechanisms of resistance are mainly present: Reduced penetration of the compound into the insect, metabolic degradation and target site insensitivity (Elbert et al. 1996).

One of the major problems in insect control is their ability to develop resistance to any chemical insecticide if used too often, thus allowing the selection of resistant genotypes. Therefore it is attractive to have many different pesticides at hand, with varying mechanisms of action. Rotating pesticides with different modes of action can limit the development of resistance, since the pest population can be treated with a second or third chemical before it builds up resistance to the first (Moffat 1993). Thus with the small percentage of imidacloprid's mode of action in commercially used insecticides (see Fig. 15 and below) the compound offers a new alternative to be included in resistance management strategies.

The number of insecticides, acaricides and nematicides acting on the nAChR had a market share in the agrochemical markets of less than 2% in 1985 (Naumann 1994). In 1990 only 1.7% of the sales volume of major insecticides relied on nAChR as a target (Leicht 1993). Since the introduction of imidacloprid to the agrochemical market the value of the nAChR as a first class target rose to over 8% (Nauen et al. 2001). Following many years of use, organophosphates, carbamates, and pyrethroids are encountering the development of insect resistance (Arther et al. 1997b). (Further information on resistance can also be found in the chapter 'Resistance'.) Similar intensive use of parasiticides has sometimes led to severe resistance against organochlorines, organophosphates, carbamates and pyrethroids as well as benzimidazoles (Kunz and Kemp 1994; Prichard 1994).

For *C. felis*, resistance was reported to a variety of insecticides, including organophosphates, carbamates, and pyrethroids (Rust 1993; Hinkle et al. 1995b). Resistance has been documented to virtually every class of insecticide ever used against the cat flea, including dichlorodiphenyl trichloroethane, dieldrin, malathion, chlorpyrifos, diazinon, propetamphos, bendiocarb, cyfluthrin, cypermethrin, fluvalinate, pyrethrin, and carbaryl (Table 9) (Lemke et al. 1989; Fox et al. 1968; El-Gazzar et al. 1986; Bossard et al. 1996), but is still controversially discussed (see chapter 'Resistance').

Even though there have been reports upon imidacloprid tolerant strains of the tobacco aphid *M. nicotianae* (Elbert et al. 1996; Devine et al. 1996), a common natural hardiness of tobacco-associated pests rather than resistance based on biochemical mechanisms was suggested (Nauen et al. 1998b). Target site insensitivity, piperonyl butoxide-sensitive oxidative metabolism or hydrolytic cleavage of imidacloprid were mainly excluded as mechanisms in low-level tolerant *M. persicae* and *M. nicotianae* (Nauen et al. 1998b). The tolerance factor between aphid strains against imidacloprid has been shown to depend strongly on the bioassay procedure used (Nauen and Elbert 1997). Nauen and Elbert (1997) could not clarify whether the low factor of tolerance of strains of *M. nicotianae* to imidacloprid in a dip test is a real mechanism-based tolerance or natural variation between closely related species. The lowered susceptibility of *M. nicotianae* or tobacco-associated *M. persicae* to chloronicotinyls could also be due to their adaptation to a plant rich in secondary plant metabolites such as alkaloids like nicotine (Nauen et al. 1998b). But the observed factors of tolerance by Nauen et al. (1998b) towards imidacloprid would never lead to a field failure at recommended application rates for *M. persicae* and *M. nicotianae*, respectively according to Dobri (1996). Elbert et al. (1996) also state that a low level of tolerance to imidacloprid's direct lethal effect does not appear to be undermining its efficacy in practice. Thus even lowered susceptibility towards *Myzus* spp. does not 'threaten' the effectiveness and usage of imidacloprid in field use.

Testing various insect pests on possible resistance to imidacloprid, Elbert et al. (1996) only reported a slightly reduced susceptibility in *M. persicae* (see above) and a resistance in an imidacloprid-selected laboratory strain of the small brown planthopper (*Laodelphax striatellus*), which did only show resistance after topical, but not after oral application.

Similar to *Myzus* species, reduced susceptibility to imidacloprid was also reported in some strains of *Bemisia tabaci* (from Spain) (Cahill et al. 1996), but it turned out to be quite low and not to influence the field performance of imidacloprid (Elbert and Nauen 1996) against these Spanish populations of *B. tabaci* (Nauen et al. 1997; Elbert and Nauen 2000).

Nevertheless in the year 2000 already registered in more than 70 countries (Elbert and Nauen 2000) imidacloprid is fulfilling the ambition of an agrochemical company in developing compounds with a new or unconventional mode of action to combat highly resistant insect pests.

Synergism of imidacloprid analogues and piperonylbutoxid in topical bioassays using houseflies suggested that they undergo oxidative detoxification (Liu et al. 1993). It is proposed that the nitromethylene carbon is the primary site of oxidative attack (in studies with 2-(nitromethylen)piperidine and related compounds) (Reed and Erlam 1978).

Prabhaker et al. (1997) demonstrated the successful selection for resistance to imidacloprid in the silverleaf whitefly *B. argentifolii* by maintaining them under continuous selection pressure. Resistance to imidacloprid developed at a slow rate with some fluctuations during the first few generations ($F_1$ to $F_{15}$, maximum resistance ratio 17-fold), followed by a gradual increase at a fairly rapid rate in the succeeding generations ($F_{16}$ to $F_{32}$, resistance ratio from 34- to 78-fold). The rise of adult resistance to insecticides often followed a predictable pattern described by Georghiou (1972) and Sawicki (1979). The rate of development of resistance is initially gradual in a population in which the resistance genes are rare and subsequently at an accelerated rate to a higher level based on the expression of resistance genes in the resistant homozygotes (Prabhaker et al. 1997; Georghiou 1972; Sawicki 1979). The slow rate at the beginning indicated a high degree of heterogeneity in response. Beginning with generation $F_{15}$ and thereafter the development indicated a more homogenous condition (Prabhaker et al. 1997). The delay in rapid increase in resistance levels to imidacloprid may be due to the low frequency of the resistance genes (Knipling and Klassen 1984). The slow and steady rise of resistance suggested the involvement of a polygenic system (Prabhaker et al. 1997).

In general an insecticide-resistant population can be derived only from individuals with pre-existing genes for resistance selected under insecticide pressure (Brown 1958). The pattern of resistance development in the studies of Prabhaker et al. (1997) is characteristic for true resistance development as reported by Hoskins and Gordon (1956). It can be summarized that even if laboratory selection experiments may not provide accurate prediction on the evolution of resistance under field conditions (Georghiou and Taylor 1976) the studies of Prabhaker et al. (1997) presented clearly the potential for development of resistance to imidacloprid in natural populations of whiteflies under field conditions. Thus the principal existence of resistance genes is demonstrated. However the laboratory selected strain lost its resistance after six generations without any further selection pressure, thus indicating tremendous fitness costs associated with the expression of the resistance genes (Nauen 2000, personal communication).

Cross-resistance in insects is a mechanism of resistance which is affecting two or three compounds regardless of the fact whether they belong to the same class of compounds or not (Elbert et al. 1996). It is a potential problem that could limit the lifetime of a new insecticide (Wen and Scott 1997). Cross-resistance between neonicotinoids (acetemiprid, thiamethoxam, and imidacloprid) has very recently been demonstrated in field strains of *B. tabaci* from Almeria, Spain (Elbert and Nauen 2000).

General guidelines to avoid or delay the development of resistance of novel insecticides are outlined (here especially for *B. tabaci* and other cotton pests) in integrated resistance management (IRM) as well as integrated pest management (IPM) programs (see Elbert et al. 1996; Horowitz and Ishaaya 1997). Considering the use of insecticides with novel modes of action general guidelines can be given and should be considered for resistance management, including long-term rotations, detailed monitoring etc.. For more information see Elbert et al. (1996) based on the IRAC guidelines (Wege 1994) and guidelines by Sanderson and Roush (1995), here for agricultural use.

Not only in farm chemicals but also in flea control on pets rotating treatment schedules and the use of different classes of insecticides or combination treatments are suggested to avoid the development of resistance (Dryden et al. 1999b).

Imidacloprid was the only adulticide out of eleven tested compounds providing 100% activity in a multi-resistant flea strain named 'cottontail', whereas fipronil only caused a reduction of 40% for two weeks. All other tested compounds did not provide control at all (Bardt and Schein 1996). In a recent study Pollmeier et al. (1999) reported a >95% efficacy of fipronil in 'cottontail' and a reference cat flea strain.

Hinkle et al. (1995b) attributed specific resistance in cat fleas to insensitive acetylcholinesterase and increased glutathione-S transferase levels (for other possible detoxification mechanisms see Hinkle et al. 1995b).

Although microsomal oxidases seem to be the most important detoxication mechanism of imidacloprid in non-responsive insects, in cat fleas detoxification probably also occurs by more than one mechanism (Richman et al. 1999).

An international group of scientists has been working since 1999 to establish a universal method for monitoring susceptibility of the cat flea to imidacloprid. Laboratory methods are in progress to be developed and standardized to determine the baseline of flea susceptibility using specified laboratory isolates as well as flea isolates from the field (Dryden and Broce 2000; Dryden et al. 2000; Blagburn et al. 2001).

# Toxicology and Pharmacology of Imidacloprid

## Symptomatology of Insects

Imidacloprid and related compounds have been tested in a variety of insect pests and proven to be very effective. The symptoms of poisoning have been described in individual species.

Whether a compound possesses a good biological activity is not only dependent on effective action at the target site but also on other parameters such as penetration, distribution and metabolism (Abbink 1991), so the method of application in toxicological studies is of importance for causing different times of onset and possibly even symptoms of poisoning.

Some insect species with their symptoms of toxicity are described below:

In American cockroaches (*Periplaneta americana*) the NMH-poisoning syndrome (after application to the ventral abdomen) is characterized by a symptomology of abdominal quivering followed, in order, by wing flexing, uncontrollable preening, leg tremors, violent whole body shaking, prostration, and finally death of the cockroach (Schroeder and Flattum 1984). Nishimura et al. (1994) observed convulsions of American cockroaches within a few minutes after injection of high doses of imidacloprid, but the symptoms did not persist for a long time. The insects either died or recovered 24 hours after injection, depending upon the dose (Nishimura et al. 1994). Depending on the dose, the symptoms of imidacloprid intoxication in the American cockroach in studies of Sone et al. (1994) were characterized by the following clear sequence of events: walking up and down (wandering) with a loss of strength in legs, leg tremor followed by a shaking of the entire body, prostration with curled abdomen and then death. Surviving insects recovered from intoxication within a day at $LD_{50}$ of imidacloprid. The symptoms clearly indicated that the insect nervous system was made hyperactive by imidacloprid (Sone et al. 1994).

Wen and Scott (1997) observed imidacloprid to be relatively fast-acting against German cockroaches (*Blattella germanica*) with the maximum effect seen two to four hours after treatment. Some cockroaches slowly recovered for up to about 72 hours, after which there was no further recovery.

The authors also observed a somewhat different effect of imidacloprid in houseflies (*Musca domestica*) which were immobilized more slowly (compared to cockroaches), with the near-maximum seen after 24 hours and no recovery (Wen and Scott 1997). In addition, treated houseflies became hyporesponsive, hyporeactive and the majority were unable to fly, but they did not develop the symptoms associated with the early signs of imidacloprid poisoning in German cockroaches (Wen and Scott 1997).

The symptomology of toxicity of injected imidacloprid in adult houseflies is also characterized by immediate convulsion with leg tremor and motion of wings, followed by paralysis (Tomizawa et al. 1995b).

Topical application on peach potato aphids (*M. persicae*) either by dipping or spraying induced neurotoxic symptoms characterized by a lack of coordination and tumbling as early symptoms, which were found to be irreversible, leading to death (Leicht 1993).

In fleas (*C. felis*) imidacloprid caused tetanic muscle contractions within minutes of exposure. Toxic symptoms manifested as intense trembling of the legs and pumping movements of the body. The nerves and muscles were constantly and irreversibly destroyed due to hyperactivity (Mehlhorn et al. 1999, 2001). Imidacloprid causes rapid loss of co-ordination of the flea with death following shortly thereafter (Bell 1997).

## Pharmacology of Imidacloprid

Numerous studies with different insect as well as vertebrate binding sites and cell structures have outlined the selective action of imidacloprid on the insect acetylcholine receptor (nAChR) and low mammalian toxicity, relying on the combination of chemical structure of imidacloprid and the molecular structure of insects nAChRs. The selective toxicity of imidacloprid was studied utilizing the nAChR preparation from various mammalian sources – rat brain (Yamamoto et al. 1995), mammalian brain (human, dog, and mouse) and chicken brain (Liu and Casida 1993), mouse N1E-115 and BC3H1 neuroblastoma cells (Zwart et al. 1994), (embryonic) rat muscle (Methfessel 1992), the *Torpedo* electric organ (Tomizawa et al. 1995a) – and insect sources - honey-bee heads (Tomizawa et al. 1995b), houseflies (Abbink 1991; Liu and Casida 1993), locust thoracic ganglion (Zwart et al. 1994), fruit fly, honey-bee, milkweed bug, German cockroach, American cockroach and cricket (Liu and Casida 1993), American cockroach fast coxal depressor motor neuron $D_f$ (Bai et al. 1991), and American cockroach sixth abdominal ganglion (Schroeder and Flattum 1984). (For detailed information on the molecular structure see chapters 'Chemical Properties of Imidacloprid' and 'Mode of Action and Direct Insecticidal Activity'.)

According to Whiting and Lindstrom (1987) over 90% of the high affinity nicotine binding sites in the mammalian brain represent $\alpha 4/\beta 2$ receptors. This type of receptor was tested as well as neuronal nAChRs containing $\alpha 7$ subunits. They showed no, respectively weak activation caused by imidacloprid and it has been generalized that the low toxicity of neonicotinoids in vertebrates was due to the insensitivity of both brain and peripheral nAChRs due to their difference in receptor type compared to insects (Yamamoto et al. 1998).

Only a number of characteristics which distinguish neonicotinoids from conventional insecticides are also low mammalian and fish toxicity (Yamamoto 1996). While nicotine is highly toxic to mammals and has limited insecticidal activities, neonicotinoids possess favorably the opposite properties (Yamamoto 1996) (Table 11). Relating mammalian toxicity to toxicity against target insects, a safety margin could be calculated and revealed high values for imidacloprid and, thus, high safety factors (Leicht 1993).

The $LD_{50}$ in most tests was greater than 2,000 mg/kg body weight (b.w.), in contrast to many currently used chemical insecticides having $LD_{50}$s of less than 50 mg/kg b.w. (Moffat 1993).

Table 11. Comparison of the toxicity of nicotine and imidacloprid

| Insecticide | $LD_{50}$ dosage in mg/kg Rat (oral) | House-fly (injection) |
|---|---|---|
| Nicotine | 53 | 272 |
| Imidacloprid | 450 | 22 |

## Pharmacokinetics and Metabolism of Imidacloprid

The pharmacokinetics of imidacloprid has been extensively studied with active ingredient (a.i.) radioactively labeled at two different molecular positions. Imidacloprid was rapidly and almost completely absorbed from the gastrointestinal tract and was extensively metabolized and rapidly excreted both following oral and intravenous administration. There were no pronounced differences between the sexes or between high and low oral doses, and there is no accumulation in specific organs (Andrews 1996a, 1996b).

The pharmacokinetics and the metabolism of imidacloprid have been studied in the rat using both methylene-$^{14}$C labeled and imidazolidine-4,5-$^{14}$C labeled a.i.. The studies showed that absorption from the intestinal tract ensued quickly and rapidly proceeded to near completion (95% in 48 hours). Rapid and complete elimination from the organism (about 96% within 48 hours) proceeded independently of the route of administration (oral or intravenous), about 75% of this by the renal route and about 21% with the feces. Most of the radioactivity eliminated by the fecal route originated from biliary excretion. There were no pronounced differences between the sexes. Whole body autoradiography did not reveal distribution to the fatty tissue, to the central nervous system or to the mineral fraction of the bones. This indicates that at a minimum dose of 20 mg/kg b.w. there was potential for accumulation and that the blood-brain barrier was fully functional.

The metabolism of imidacloprid has been studied in rats treated with methylene-$^{14}$C- or imidazolidine-4,5-$^{14}$C labeled a.i.. Two major metabolic routes exist in mammals. The first is the oxidative cleavage to imidazoline and 6-chloronicotinic acid. Imidazoline is excreted directly in the urine, the other moiety is degraded to mercapturic and then methylmercaptonicotinic acid. Quantitatively, this route only accounts for 7% of the total biotransformation, thus deficiencies in conjugation capacities will be without toxicological consequences. The second route is hydroxylation of the imidazoline ring followed by elimination of water and the formation of an unsaturated metabolite.

The proposed metabolic pathway of imidacloprid in the rat is shown in Fig. 16. Essentially similar metabolic patterns have also been determined in two other species, the goat and the hen.

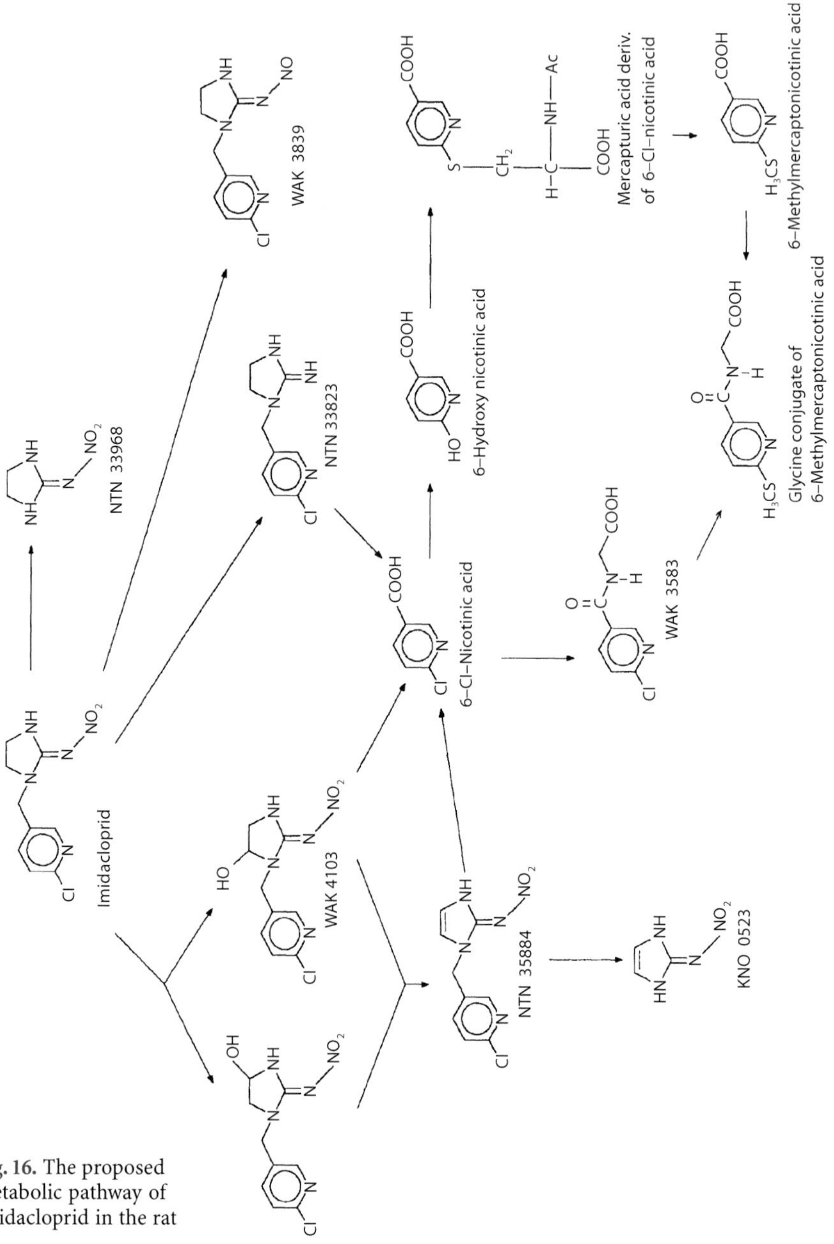

Fig. 16. The proposed metabolic pathway of imidacloprid in the rat

# Toxicology of Imidacloprid

The functional selectivity (see chapter 'Mode of Action and Direct Insecticidal Activity') and the favorable pharmacokinetic parameters (see chapter 'Phar-macokinetics and Metabolism of Imidacloprid') of imidacloprid are the basis for its low and favorable toxicity in mammals. Reference to the large number of toxicological studies is not made individually as all toxicological information has been accumulated in toxicological expert opinions (Andrews 1996a, 1996b). These assessments and all individual study reports have been submitted to and accepted by international and national regulatory agencies. A short description of the toxicology and metabolism of imidacloprid has recently been published (Thyssen and Machemer 1999).

### Single-dose toxicity

In rats and mice, imidacloprid was of moderate acute oral toxicity and, in rats, of moderate to low acute intraperitoneal toxicity. Imidacloprid was untoxic to rats when it was inhaled at maximal technically attainable dust or aerosol concentrations. Important for the use as veterinary spot-on preparation is the very low acute dermal toxicity. The $LD_{50}$ levels obtained in the acute studies with different animal species exposed by several routes are shown in Table 12. Toxic signs were observed following high doses of imidacloprid, were reversible within six days and consisted of behavioral, respiratory and motility disturbances, narrowed palpebral fissures, and transient trembling and spasms.

From the known very low lipophilicity of imidacloprid and also from the difference between intraperitoneal (150 mg/kg b.w.) and dermal (5,000 mg/kg b.w.) $LD_{50}$ values in the rat it can be concluded that transcutaneous absorption will be insignificant.

Local tolerance was very good in tests with rabbits. In rabbits, there was neither irritation to the skin nor to the mucosa of the eye. Imidacloprid did not give rise to skin sensitization in guinea pigs.

When given together with other insecticides there were no potentiating effects in acute oral studies. This has been shown specifically for the combination of imidacloprid with the pyrethroid cyfluthrin and with the organic phosphorous ester methamidophos. The absence of potentiating effects of mixtures of imidacloprid and other insecticides was not an unexpected result in the light of the novel mechanism of action of imidacloprid.

### Repeated-dose toxicity

The effect of repeated administration of imidacloprid has been studied in rats, dogs and rabbits that were treated daily for periods ranging in length from four weeks to two years.

Most appropriate for risk assessment for the use of imidacloprid as a veterinary spot-on product is the dermal route of administration. A three-week dermal study on rabbits resulted in a NOEL (no-observed-effect level) for imidacloprid of 1000 mg/kg b.w./day, i.e. some 100- to 25-fold higher than the recommended single dose in the target animal species (Table 13).

Following oral treatment in rats a NOEL of 14.9/83.3 mg/kg b.w./day (males/females) was determined in a 13-week feeding study. An additional two-year chronic feeding rat

Table 12. Acute toxicity of imidacloprid a.i.

| Application route/species | Vehicle | Sex | LD$_{50}$ (mg/kg b.w.) |
|---|---|---|---|
| **per os** | | | |
| Rat | cremophor EL | m | 424 |
| | /water | f | 450-475 |
| | cremophor EL | m | 642 |
| | /water | f | 648 |
| | cremophor EL | m | 504 |
| | /water | f | 379 |
| Mouse | cremophor EL | m | 131 |
| | /water | f | 168 |
| **dermal** | | | |
| Rat | NaCl solution | m, f | > 5,000 |
| **inhalation** | | | |
| Rat | 1 x 4 h, dust | m, f | > 5,323 mg/m³ air * |
| | 1 x 4 h, aerosol | m, f | > 69 mg/m³ air * |
| **intraperitoneal** | | | |
| Rat | cremophor EL | m | 160 – 170 |
| | /0.9% NaCl | f | 186 |

m = male; f = female
* = maximal technically obtainable and analytically determined concentration

Table 13. Repeated-dose toxicity of imidacloprid a.i.

| Species | Route of administration | Duration of exposure | NOEL (male/female) |
|---|---|---|---|
| Rabbit | dermal | 15 days | 1,000 mg/kg b.w./day |
| Rat | dust inhalation | 4 weeks | 5.5 mg/m³ air/day |
| Rat | medicated feed | 13 weeks | 14.0/83.3 mg/kg b.w./day |
| Rat | medicated feed | 13 week (neurotoxicity) | 196 mg/kg b.w./day |
| Dog | medicated feed | 13 weeks | 15 mg/kg b.w./day |
| Dog | medicated feed | 52 weeks | 15 mg/kg b.w./day |
| Mouse | medicated feed | 52 weeks | 65.6/103.6 mg/kg b.w./day |
| Rat | medicated feed | 104 weeks | 5.7/24.9 mg/kg b.w./day |
| Rat | gavage | day 6-15 of gestation | dams: 10 mg/kg b.w./day pups: 30 mg/kg b.w./day |
| Rabbit | gavage | day 6-18 of gestation | dams: 8 mg/kg b.w./day pups: 24 mg/kg b.w./day |
| Rat | medicated feed | 2 generations | parents: 12.5/6.7 mg/kg b.w./day reproduction: 12.5 mg/kg b.w./day |

study yielded NOELs of 5.7/24.9 mg/kg b.w./day (males/females). The long-term feeding studies with rats and mice showed that imidacloprid is not carcinogenic. The liver has been identified as the main target organ in all repeated-dose studies. Slight increases in liver weight and increases in cytochrome P450 and its dependent enzymes were seen. In male rats, cell hypertrophy and sporadically cellular necrosis were observed. These alterations were reversible high-dose phenomena and were combined with up to 10% reduction in body weight gain. A secondary effect which was only observed in high-dose male rats was degeneration of the testicular bulbus and a slight decrease in blood clotting time.

The repeated-dose studies also showed that imidacloprid does not effect the immune system in any untoward manner.

The absence of neurotoxic effects has been shown in a special 13-week feeding study with rats. Functional observational battery, grip strength, foot splay, motor and locomotor activity in a figure-eight maze, habituation and a special pathology with emphasis on neuronal tissue were performed. The NOEL for neurotoxicity was 196 mg/kg b.w./day in the 13-week study and 307 mg/kg b.w. in a preceding acute neurotoxicity study.

Repeated-dose inhalation studies in rats under dynamic conditions for four weeks (six hours/day; five days/week) revealed a NOEL of 5.5 mg/m$^3$ air, equal to about 2.4 mg/kg b.w./day.

Oral administration to dogs, one of the target animal species, for 13 weeks or for 1 year resulted in a NOEL of 15 mg/kg b.w./day.

### Reproductive toxicology including teratogenicity

Reproductive toxicity has been tested in rats and rabbits and no potential to cause primary embryotoxicity or teratogenic effects has been discovered. The NOEL for maternal and fetal toxicity tested on rabbits were 8 mg/kg b.w./day and 24 mg/kg b.w./day, respectively. In rats, the respective NOELs were 10 mg/kg b.w./day and 30 mg/kg b.w./day. No primary toxic effects on reproduction were discovered in a 2-generation rat study which resulted in NOELs for parent animals of 12.5/6.7 mg/kg b.w./day (males/females) and a NOEL for reproduction of 12.5 mg/kg b.w./day.

### Mutagenicity and genotoxicity

The test systems employed to exclude mutagenic and genotoxic effects ranged from bacteria and isolated mammalian cells to mammals in vivo and included tests for point mutations, for chromosome aberrations and for DNA damage. All tests with imidacloprid were negative, while the expected responses were obtained with the respective positive control substances. It had to be concluded that imidacloprid possesses neither a mutagenic nor a genotoxic potential.

## Toxicology of Imidacloprid 10% Spot-on

Acute toxicity tests and tests for local tolerability with imidacloprid 10% spot-on (Advantage®, Bayer AG) have been conducted to assess whether contact with the product may give rise to untoward effects on the user of the product, on the pet owner handling

Table 14. Acute toxicity of imidacloprid 10% spot-on

| Application route/species | Sex | LD$_{50}$ (mg/kg b.w.) |
|---|---|---|
| **per os** | | |
| Rat | male | 1,943 |
| | female | 1,732 |
| **dermal** | | |
| Rat | male | > 2,000 |
| | female | > 2,000 |
| **inhalation (1 x 4 hours)** | | |
| Rat | | > 2,415 mg/m³ air * |
| **irritation** | | |
| Skin irritation | female | mildly irritating |
| Eye irritation | female | moderately irritating |
| **sensitization** | | |
| Guinea pig | female | negative |

* = analytically determined concentration

treated pets and also on treated dogs and cats. The studies conducted with imidacloprid 10% spot-on solution reflect the results obtained for the a.i. (see chapter 'Toxicology of Imidacloprid' ). Very low acute oral, dermal and inhalation toxicity was obtained for rats and no skin sensitizing effects in guinea pigs could be observed. The results of the individual studies are shown in Table 14.

Treatment with imidacloprid 10% spot-on  gave rise to mild irritation of the conjunctival mucosa with corneal opacity, iritis, conjunctival redness and chemosis for two weeks and mild irritation and erythema of the skin. These effects have supposedly been caused by the solvent, as pure imidacloprid was neither irritating to the eye nor to the skin.

Signs of skin irritation and ocular exposure, occurrence of which would easily be noted, have not been encountered however in the clinical efficacy and tolerability studies. Prophylactically, one should avoid contaminating the eyes or snout of animals whilst treating them.

## Tolerability of Imidacloprid 10% Spot-on in Dogs and Cats

The recommended single dermal dose of 10 mg imidacloprid/kg b.w. which is administered to dogs and cats in the form of imidacloprid 10% spot-on is of the same order of magnitude as the oral NOELs obtained in long-term feeding studies and in the teratogenicity and reproductive toxicity studies (see chapter 'Toxicology of Imidacloprid'). Especially if it is taken into account that only an insignificant fraction of the dermally administered a.i. can be assumed to be absorbed transcutaneously, one can conclude that tolerability problems were unlikely to arise in dogs and cats that are occasionally and dermally treated with imidacloprid 10% spot-on.

Nevertheless, tolerability studies with imidacloprid 10% spot-on have been conducted in dogs and cats in order to establish that even multiples of the recommended dose are well tolerated.

Treatment with imidacloprid 10% spot-on (at two-fold the recommended dose rate and repeated after two weeks instead of the recommended four weeks) has been shown to be compatible with the recommended dose rates of commonly used canine remedies (fenthion (Tiguvon®, Pro Spot®), lufenuron (Program®), milbemycin (Interceptor®) and a combination product containing febantel, pyrantel and praziquantel (Drontal® Plus)) and feline remedies (lufenuron (Program®) and a combination product containing pyrantel and praziquantel (Drontal®)).

Individual treatment with exact dosing under the control of the person applying the product, treatment at a site that is difficult to lick or reach with the paws, a high safety margin of the a.i., its rapid metabolism and excretion after ingestion and a bitter taste of the product which counteracts spontaneous oral ingestion all work together to achieve a very good tolerability in the target animal species. The safety and the good tolerability of the commercial formulation for dogs and cats are well documented and toxicity studies in target animal species revealed no signs of local or systemic intolerance when the 10% imidacloprid formulation was applied dermally. Finally, imidacloprid 10% spot-on was considered safe also in gravid lactating bitches and queens.

Oral contact with the product may give rise to untoward side effects in young or very low weight animals in case they are overdosed. In larger or adult animals, oral contact is inconsequential. Even if 1/10 of the dermal dose (equivalent to 1-2.5 mg imidacloprid/kg b.w.) were ingested through licking and even if this occured repeatedly, this could not be expected to cause any adverse reactions. Dermal treatment in special situations (e.g. treatment of skin lesions and possibly increased transcutaneous resorption) will not put the treated animal species at risk.

Puppies younger than eight weeks and kittens younger than ten weeks should not be treated, although there is no indication that this would entail a specific risk. Provided the standard instructions for handling the product are followed, the use of the veterinary medicinal product 10% spot-on does neither entail undue risks for the target animal species cat and dog, nor for the environment nor for man (see also chapters 'Ecological Effects and Ecotoxicological Studies' and 'Safety Assessment for Humans').

Results obtained in studies employing higher than recommended doses (two to ten times) of the veterinary medicinal product imidacloprid 10% spot-on  and also multiple treatments at shorter intervals than recommended, confirm both the safety and the good tolerability of the commercial formulation for dogs and cats. In practice, application by the pet owner of such high doses of the product are unlikely to occur as excess product will then run off.

Detailed examination of the tolerability of imidacloprid 10% spot-on  in the two main pet species have been conducted involving different types and ages of dogs and cats. The results are listed below. The doses are given in mg imidacloprid/kg b.w. although imidacloprid 10% spot-on was used in the tolerability studies described.

**Tolerability in dogs**

Acute overdosing of 10-day old puppies was tested following dermal application of 200 mg imidacloprid/kg b.w. (20-times the recommended dose). No side effects were recorded.

Acute overdosing of adult dogs was studied by administration of ten-fold the recommended dose (100 mg imidacloprid/kg b.w.). This treatment was tolerated by ten dogs without clinical signs occurring during a 72 hours post observation period. Dermal application of 200 mg imidacloprid/kg b.w. (20-times the recommended dose) did also not cause any side effects.

The effect of repeated overdosing was tested in growing dogs. Dermal application of 10 mg imidacloprid/kg b.w. (recommended dose) or of 50 mg imidacloprid/kg b.w. (five times the recommended dose) at seven day intervals for eight consecutive weeks did not cause any side effects in dogs as young as eight weeks when first treated.

The effect of repeated overdosing was also tested on adult dogs, gravid and lactating bitches and greyhounds (Table 15).

Daily dermal application of 50 mg imidacloprid/kg b.w. (five times the recommended dose) for three consecutive days did not cause any side effects. Dermal application to gravid bitches of 10 to 30 mg imidacloprid/kg b.w. (up to three times the recommended dose) twice, two to four weeks apart, did not cause any side effects at various stages of pregnancy. Dermal application to lactating bitches of 18 to 30 mg imidacloprid/kg b.w. (up to three times the recommended dose) twice, three to four weeks apart, as well as double treatment four weeks apart at a rate of 20 mg imidacloprid/kg b.w. imidacloprid (two times the recommended dose) did not cause any side effects at various stages of lactation in bitches some of which had also been treated during pregnancy.

Greyhounds, a very sensitive breed, were dermally treated with 40 mg imidacloprid/kg b.w. (four times the recommended dose) twice two weeks apart. No adverse reactions were observed.

The results of application of high doses of imidacloprid 10% spot-on in ten-day-old and less than 1 kg weight puppies as well as in pregnant and lactating bitches confirm that adverse effects - general as well as local reactions - are not to be expected even after overdosing five- to ten times overdosing or when treating at shorter treatment intervals than recommended.

**Tolerability in cats**

Acute overdosing of weaned kittens was tested following dermal application of 80 mg imidacloprid/kg b.w. (eight times the recommended dose). This treatment did neither result in adverse affects in 12-13-week-old kittens nor did treatment with 240 mg imidacloprid/kg b.w. (24-times the recommended dose) of 6-10-week-old kittens cause such effects. Single treatment at eight-fold the recommended dose (80 mg imidacloprid/kg b.w.) was tolerated by weaned kittens without clinical signs, either local or systemic, at four and 24 hours and again at seven and 14 days post treatment. The minimum weight of the treated kittens was 1.1 kg (Table 16). One kitten showed several rapid lip licking movements seven minutes after treatment, a sign considered to be an unspecific reaction sporadically occurring in highly taste- or smell-sensitive cats rather than a sign of intoxication.

Table 15. Treatment schedule of six bitches with imidacloprid 10% spot-on during pregnancy and lactation

| Dog breed | Body weight (kg) | Gestation time, treatment day | Dosage (mg/kg) | Lactation time, treatment day | Dosage (mg/kg) | Puppies born alive | Observation period p.p.(weeks) |
|---|---|---|---|---|---|---|---|
| Bull Terrier | 22.8 | 23, 51 | 20, 20 | 3, 31 | 20, 20 | 8 | 7 |
| Rottweiler | 31.6 | 27, 53 | 20,20 | -, - | -, - | 3 | 1 |
| Rhodesian Ridgeback cross | 21.5 | -, - | -, - | 2, 30 | 20, 20 | 6 | 7 |
| Bull Terrier | 21.9 | 39, 52 | 20, 30 | 4, 33 | 30, 30 | 9 | 5 |
| Chihuahua | 4.7 | 22, 47 | 20, 20 | 1, 21 | 20, 20 | 3 | 7 |
| Maltese Terrier | 5.5 | 28, 56 | 10, 10 | 10, 38 | 19, 18 | 4 | 39 |

p.p. = post partum

Table 16. Dose rates of imidacloprid given in an overdose trial in kittens

| No. of kitten treated | Starting body weight (kg) | Age (weeks) | Dosage (mg/kg) | Dose volume (ml) |
|---|---|---|---|---|
| 7 | 1.09 - 1.47 | 12 - 13 | 80, once | 0.87 - 1.18 |
| 8 | 0.69 - 0.91 | 6 - 10 | 240, once | 1.64 - 2.18 |
| 4 | 0.51 - 0.67 | 6 | 180 - 239, twice | 1.2 |

Acute overdosing of adult cats following a single treatment at ten-fold the recommended dose (100 mg imidacloprid/kg b.w.) was tolerated in ten cats without clinical signs during a seven day post observation period.

The effect of repeated overdosing of 6-week-old kittens was assessed by dermal application of 180-239 mg imidacloprid/kg b.w. (~18- to 24-times the recommended dose) twice (two weeks apart), a treatment that did not cause any side effects (Table 16). Dermal application of five-fold the recommended dose of imidacloprid spot-on 10% (2 ml solution) to initially 6-week-old weaned kittens at weekly intervals (i.e. four times shorter treatment interval than recommended) for eight subsequent weeks resulted in negligible clinical signs and generally good tolerability. Again, oral contact with the test substance which occurred in naïve cats via licking off the paws subsequent to scratching the application site was observed up to 20 minutes after the first and up to ten minutes after the second treatment, but never thereafter. Licking induced transient salivation (and often ingestion of feed, so as to cleanse the mouth of some unpleasant taste). No evidence of treatment-related effects was observed in growing cats, initially as young as ten weeks, treated dermally with 10 or 50 mg imidacloprid/kg b.w. at seven day intervals for eight weeks.

The effect of repeated overdosing was also tested in adult cats and gravid and lactating queens. Daily dermal application of 50 mg imidacloprid/kg b.w. imidacloprid (five times the recommended dose) for three consecutive days did not cause any side effects, nor were treatment-related effects detected in hematological and clinical chemical determinations conducted one week after the first treatment. One animal in one of the trials which managed to lick part of the large volume applied and running down the neck showed transient salivation and exhibited sneezing and unilateral nasal discharge for one day after treatment. No other abnormalities were observed and the signs resolved by the 9th day of clinical examination. This observation stresses the importance of avoiding any direct or indirect (through licking of paws) oral contact with the product.

Dermal application of 40 mg imidacloprid/kg b.w. (four times the recommended dose) twice during gestation or three times during lactation did not cause any side effects in either queens or kittens.

Generally, the dermal application of imidacloprid 10% spot-on has a large safety margin in cats. An important prerequisite to achieve good tolerability and to prevent any untoward reversible effects is the exclusively dermal application and the strict impeding of any form of oral contact (e.g. ingestion by licking).

## Ecological Effects and Ecotoxicological Studies

Studies on the effects of wildlife and aquatic as well as terrestrial non-target organisms were performed for the registration of imidacloprid as an agrochemical. The overall risk assessment indicates that the use of imidacloprid in turf and agricultural crops will result in minimal risk to wildlife (Mullins 1993). For detailed information see also chapter 'Pharmacology of Imidacloprid'. Although imidacloprid is highly effective against a broad spectrum of pests, it is unusually safe in environmental terms (Pflüger and Schmuck 1991).

Residue studies indicate that imidacloprid should not leak into groundwater under field conditions. Further laboratory studies showed that imidacloprid in water is rapidly degraded by sunlight with a water column half-life in outdoor aquatic dissipation studies of approximately 1.4 days (Mullins 1993).

Considering the original indication of imidacloprid as insecticide used to control numerous insect pests of the agricultural sector, intensive studies were performed testing the toxicity against soil microbes, earthworm bees, ground beetles (Carabidae), beneficial insects (ladybirds, syrphid flies, lacewings, hymenoptera, diptera and rove beetles), and predatory mites. The harmful effect of imidacloprid was slight at most (beneficial insects) or transient (earthworms in spray application four times the standard dose). Solely against bees the compound was hazardous, giving the advise to use imidacloprid outside the blossom time. In tests with algae, water fleas, and fish the safety margins were so large that the possibility of a harmful effect on these aquatic creatures can be ruled out. Soil organisms, algae, water fleas and various fish species are not endangered even if imidacloprid is used at field rates several times higher than recommended (Pflüger and Schmuck 1991). Laboratory testing on the standard array of aquatic species indicates unusually low toxicity for an insecticide. $LC_{50}$ values for fish range from >105 to 211 ppm and for shrimp (being considerably more sensitive) at 0.034 ppm respectively in the same range for other freshwater invertebrates (amphipods and chironomids). An outdoor microcosm study evaluating the potential effects on invertebrate communities showed temporary reductions in specific aquatic invertebrates at relevant environmental concentrations, but also a rapid recovery due to the short half-life of imidacloprid in the water column (Mullins 1993) (Table 17).

Table 17. Selection of toxicological data in mammals, birds, fish and invertebrates

| Organism | Toxicological test | Dosage (mg/kg) |
|---|---|---|
| **Birds** | | |
| Canary | $LD_{50}$ | 25 - 50 mg/kg b.w. |
| Pigeon | $LD_{50}$ | 25 - 50 mg/kg b.w. |
| Japanese quail | $LD_{50}$ | 31 mg/kg b.w. |
| Virginia quail, 5 days feeding | $LC_{50}$ | 1420 mg/kg foodstuff |
| Virginia quail, reproduction toxicity | NOEC | > 243 mg/kg foodstuff |
| **Fish** | | |
| Rainbow trout | $LC_{50}$, 96 hours | 237 mg/l |
|  | threshold (21 days, 15° C) | 29 - 62 mg/l |
| Carp | $LC_{50}$, 96 hours | 280 mg/l |
| **Invertebrates** | | |
| Earthworm (*Eisenia fetida*) | $LC_{50}$, 14 days | 10.7 mg/kg |
| Water flea (*Daphnia magna*) | $EC_{50}$, 48 hours | 85 mg/l |
|  | reproduction toxicity threshold, 21 days | 1.8 - 3.6 mg/l |
| Algae growth (*Scenedesmus*) | $EC_{50}$, 96 hours | > 10 mg/l |

Imidacloprid was also tested on mammals and birds which possess potential risk of contact when the compound is used as farm chemical. The $LD_{50}$ for birds ranges between 25-50 mg/kg which can be regarded as acutely toxic (Pflüger and Schmuck 1991) (Table 17). The subacute toxicity, a more important factor for judging residues in wild animal nutrition, is low with a $LD_{50}$ of >1,000 ppm (Toll 1990b, 1990c). Negative effects on reproduction are not expected with NOEC-values of 125-243 ppm in trials of ducks and quails (Toll 1990a, 1990b). Birds have been shown to avoid treated seeds due to a repellent effect (Naumann 1994).

Trials examining the effects of spraying of imidacloprid have not shown any risk for vertebrates even if they were to feed exclusively on vegetation with maximum load (Pflüger and Schmuck 1991).

Trials with treated seeds resulted in definite symptoms in vertebrates shortly after consumption from which test animals recovered completely within a short time at low dosage (2.5-5 mg/kg). Generally a risk for wild animals cannot be excluded at first, but a strong repellent effect will prevent toxic quantities from being ingested in practice even if imidacloprid is applied in a seed dressing or in granules. Therefore a risk for mammals is not expected (Pflüger and Schmuck 1991). All in all imidacloprid has favorable mammalian safety characteristics (Werner et al. 1995a; Jacobs et al. 1997a).

Summarizing, laboratory and field tests conducted on aquatic and terrestrial non-target organisms have been performed and the overall risk assessment indicates that the use of imidacloprid in turf and agricultural crops will result in minimal risk to wildlife (Mullins 1993).

Concerning imidacloprid as flea adulticide (imidacloprid 10% spot-on ) the pipette in which the product is sold is completely emptied when application is made correctly. The product is intended for individual use in companion animals (cats and dogs), the amount of imidacloprid applied per treatment is small, and the risk of running off is minimal, therefore it is concluded that imidacloprid 10% spot-on  does not present any ecotoxicological risk (Andrews 1996a, 1996b). Following treatment of companion animals imidacloprid is incorporated into the lipid layer of the skin and on the hair (Mehlhorn et al. 1999), thus further reducing the possibility of contaminating the ecosystem. Nevertheless, recently treated dogs should not go swimming, be bathed or get soaking wet by rain, lest a fraction of the a.i. just applied could be washed off which might be associated with a reduction of insecticidal effectiveness.

## Safety Assessment for Humans

There is no experience with imidacloprid poisoning in man. No anomalies were observed at preventive occupational medical examinations of employees who could have come into contact with imidacloprid during production of the a.i. or during formulation and packaging of products containing imidacloprid. There is no specific antidote for imidacloprid. Dosing with activated charcoal should be initiated in suspected cases of poisoning with imidacloprid, as this treatment has been shown to alleviate toxic signs in intoxicated laboratory animals.

Normal hygienic measures are advised by the manufacturer. These are in detail not to eat, drink or smoke during application of imidacloprid 10% spot-on and to wash hands thoroughly after treating animals. There is no reason why wearing gloves is required. Any spillage of the product on human skin should be removed by washing, and eyes should be flushed with water. As a general precaution, veterinary medicinal and insecticidal products should be kept out of the reach of children although the delivery system of imidacloprid 10% spot-on is childproof.

Man may be exposed to imidacloprid inadvertently through consumption of agricultural crops that have been protected against insects by imidacloprid-containing insecticidal formulations, when applying imidacloprid to plants or animals and when handling animals treated with imidacloprid 10% spot-on. Although toxicological studies have demonstrated that imidacloprid has a favorable toxicological profile (see chapter 'Toxicology of Imidacloprid'), quantitative assessments of human safety have been made with respect to different uses of imidacloprid.

The significance of residues and other exposure from the agricultural use of imidacloprid has been assessed (Mullins 1993) and large margins of safety have been calculated. A Chronic Dietary Exposure Analysis (Tas 1990) estimating the mean chronic exposure to constituents in foods comprising the diets of the average US population and 22 population subgroups indicated that there will be a large measure of safety to the consumer relative to potential residue amounts of imidacloprid in food, given the conservative nature of the analysis and the large margin of safety associated with the proposed reference dose (RfD) (relying on the low toxicity of imidacloprid to mammals). Similar favorable results exist for margins of safety (MOS) calculated for inhalation and dermal exposure of mixers, loaders and applicators. The large MOS suggest that imidacloprid can be used safely, without significant risk to people who mix, load or apply it.

The recommended dosage of imidacloprid for flea treatment of pet animals is 10 mg/kg b.w. via the dermal route of application, which is achieved by administration of 0.1 ml/kg b.w. of the formulated product imidacloprid 10% spot-on at a treatment interval of four weeks (Heeschen 1995; Liebisch and Heeschen 1997). The a.i. has distributed over the entire body surface already after twelve hours post applicationem and remains detectable in considerable amounts on the skin and in the hair coat of the treated animal (Fichtel 1998). Imidacloprid is included in the lipid layer of the skin surface and the hair, thus reducing the amount of imidacloprid available for skin penetration (Mehlhorn et al. 1999).

When being used as a veterinary medicinal preparation, exposure of pet owners, pet care takers, professional groomers or the veterinary professionals to imidacloprid 10% spot-on has to be considered.

The most critical exposure situation arises when children have close contact with treated pet animals. Being lighter in weight than adults their potential per kilogram exposure will be higher than that of similarly exposed adults. Therefore, a detailed assessment of the potential risk that may result from physical contact between children and pet animals treated with imidacloprid 10% spot-on was performed on the basis of results of a stroking test. The following of worst case scenario parameters have been assumed and used as basis for the quantitative risk assessment (Andrews and Bomann 1996):

1. The dog receives the highest dose rate possible according to label directions (500 mg imidacloprid/dog).
2. The body weight of the child who could have contact with the treated animal is 10 kg.
3. The size of the child's hand is about 1/4 the size of an adult hand. Assuming that both hands contact the treated animal, in total half the amount of a.i. transferable to an absorbent cotton glove worn by an adult during the stroking test is the maximal amount transferred to both hands of the child.
4. The maximal amount of a.i. that may be ingested orally is equal to the amount transferred to the skin of the hands.

The maximum amount of imidacloprid transferable from hair and skin of the treated body area of dogs to the skin of man was estimated by determination of the amount of imidacloprid incorporated into absorbent cotton gloves during stroking of a treated dog. The treated area was stroked 30-times with the hand wearing a cotton glove, followed by extraction of the glove and analytical determination of imidacloprid. The amounts of imidacloprid transferred during stroking decreased rapidly with time (Table 18).

Two separate scenarios were evaluated:

1. The first scenario considers the risk associated with a single contact with the treated animal directly after treatment. Basis for the assessment is the amount of the commercial product, imidacloprid 10% spot-on, transferred from the still wet treatment area to the absorbent cotton glove ten minutes after treatment of the animal.
2. The second scenario considers the risk associated with repeated contact with treated animals. Basis for the assessment is the amount of a.i., imidacloprid, transferred from the dry treatment area to the absorbent cotton glove 24 hours after treating the animal.

Comparisons of the level of human exposure calculated for both scenarios with the appropriate doses of either imidacloprid 10% spot-on or of imidacloprid a.i. tolerated by laboratory animals without effect result in margins of safety. The type of exposure

Table 18. Results of stroking tests with dogs treated with imidacloprid 10% spot-on

| Sampling time p.t. | Maximal amount of imidacloprid transferred to absorbent cotton glove (mg/glove) | Maximal amount imidacloprid transferable to hands of children (mg/10 kg child) | Maximal imidacloprid exposure (mg/kg child) |
|---|---|---|---|
| 10 min | 24.9 | 12.45 | 1.245 |
| 1 h | 17.3 | 8.65 | 0.865 |
| 12 h | 3.9 | 1.95 | 0.195 |
| 24 h | 2.7 | 1.35 | 0.135 |

considered here is comparable to the occupational exposure, as it might also be repeated but certainly will not be life-long. Safety margins of 10 which are usually considered to be sufficient for the setting of permissible levels for occupational exposure (Zielhuis and Van Den Kerk 1978) were also considered sufficient here.

The appropriate toxicological data that have to be used for the calculation of the safety margins are shown in Table 19. The NOEL obtained in acute toxicity studies with imidacloprid 10% spot-on have to be used for scenario 1, while data obtained in repeated-dose studies with imidacloprid have to be used for scenario 2.

A single, incidental dermal exposure of a child (scenario 1) may result in a maximal dermal exposure to 12.45 mg imidacloprid, which is equivalent to a dermal dose of 1.25 mg imidacloprid/kg b.w. (Table 18, Table 20). This dose was compared with the dose of imidacloprid a.i. which had been tolerated without effects by rats in the acute dermal study (5,000 mg/kg b.w.). Thus the maximum dermal dose to which a child may be exposed in a single exposure shortly after treatment is 4,000-fold less than the rat dose, indicating a margin of safety of 4,000. In repeated dermal exposure (scenario 2) it was

Table 19. Relevant toxicological data used for the calculation of safety margins

| Type of test | Toxicological endpoint | Dose |
|---|---|---|
| **Results of acute studies with imidacloprid 10% spot-on** | | |
| acute dermal (rat) | $LD_{50}$ | > 2000 mg/kg b.w. |
| | lowest-effect level | > 2000 mg/kg b.w. |
| | no-observed-effect level | 2000 mg/kg b.w. |
| acute oral (rat) | $LD_{50}$ | 1742 mg/kg b.w. |
| | lowest-effect level | 1020 mg/kg b.w. |
| | no-observed-effect level | 600 mg/kg b.w. |
| **Results of repeated-dose studies with imidacloprid a.i.** | | |
| subacute dermal | lowest-effect level | > 1000 mg/kg b.w./day |
| (3-week, rabbit) | no-observed-effect level | 1000 mg/kg b.w./day |
| subchronic oral | lowest-effect level | 56 mg/kg b.w./day |
| (13-week, rat) | no-observed-effect level | 14 mg/kg b.w./day |

Table 20. Exposure scenarios for imidacloprid and their margins of safety

| Type of exposure | Maximal exposure of children (mg/kg b.w.) | | Minimal dose tolerated without effect by laboratory animals (mg/kg b.w.) | | Margin of safety |
|---|---|---|---|---|---|
| single dermal contact | 12.5 | * | 2000 | * | 160 |
| repeated dermal contact | 0.135 | # | 1000 | # | 7,407 |
| single oral contact | 12.5 | * | 600 | * | 480 |
| repeated oral contact | 0.135 | # | 14 | # | 104 |

\* = mg imidacloprid 10% spot-on (commercial product)/kg b.w.
# = mg imidacloprid a.i./kg b.w.

assumed that a child may be exposed to a maximum daily dose of 0.135 mg imidacloprid/ kg b.w.. This dose was compared with the daily dose tolerated without effects by rabbits in a 3-week dermal study (1,000 mg/kg b.w./day). The maximum daily dose in a child was 7,407-fold less than the NOEL in the rabbit, indicating a margin of safety of 7,407.

The potential risk of oral intake in humans exists, especially for children, so they should not be allowed to apply the compound and stroke the treated animals directly after application which could cause contaminated fingers and thus make an oral exposure possible, however, oral exposure is considered not to be very likely to occur. The maximum oral intake was assumed to be equal to the maximum dermal exposure, although the bitter taste of imidacloprid 10% spot-on makes it questionable whether the oral uptake will really be that high.

The maximum amount of imidacloprid taken up orally was considered to be 1.25 mg/kg b.w. after single incidental exposure (scenario 1). Compared with the NOEL of rats in the acute oral study (600 mg/kg b.w.), the maximum oral dose to which a child may be exposed is 480-fold less than the rat dose, indicating a margin of safety of 480. In the case of repeated oral exposure (scenario 2), a maximal oral dose of 0.135 mg imidacloprid/kg b.w. was compared to the NOEL of rats from the subchronic feeding study (14 mg/kg b.w./day), resulting in a 104-fold less human oral exposure dosage, and indicating a margin of safety of 104.

Both assumptions (scenarios 1 and 2) were undoubtedly worst-case assumptions, as it was shown that the amount transferred by stroking rapidly decreased with time after treatment. Nevertheless, all comparisons of dermal and oral exposure data of a child with doses that have not caused adverse reactions in laboratory animals result in high margins of safety (see Table 20) even though the calculations were based on worst-case assumptions.

It can be concluded that even in the theoretical and most unfavorable situation of close physical contact between a child and a recently treated animal it can be stated that children are not put at risk, regardless of whether single dermal, repeated dermal, single oral, repeated oral, or combinations of dermal and oral exposures occur. The calculated margins of safety ranged from 104 to 7,407 and were thus considerably higher than 10.

# Imidacloprid as a Veterinary Product for Flea Control

For the treatment of fleas in companion animals the selection of an insecticide and its formulation should generally be based upon the species and age of the animal to be treated, the level of infestation, the rate of potential reinfestation, the thoroughness of environmental treatment, and the insecticide resistance pattern of the flea population (Rust and Dryden 1997). However, in most cases insecticide resistance patterns are unknown, and ultimate selection of an insecticide formulation is actually based on economics and ease of product (Rust and Dryden 1997). Another influence in the choice of the ideal insecticidal treatment of fleas is the increased pet owner consideration of safety and toxicology.

Until the early nineties, imidacloprid was mainly used in crop protection, at which time its activity against fleas was recognized in flea-infested dogs by Dr. Terry Hopkins, Bahrs Hill Research Station, Beenleigh, Qld, Australia. Worldwide development was initiated, resulting in Advantage® (imidacloprid 10% spot-on) topical solution being first registered in the USA in March 1996.

Imidacloprid applied to glass was tested against adult cat fleas and found to be more active than most other insecticides in literature (Richman et al. 1999). The picture was the same for larval cat fleas tested on glass (Richman et al. 1999). Both methods were compared with the results of several studies testing different organophosphates, carbamates, and pyrethroids (see Richman et al. 1999 for further publications).

In Advantage®, imidacloprid is offered in a clear to slightly yellow colored solution with a slight aromatic odor, containing 9.1% (wt/wt) (10% w/v) imidacloprid. The criteria for the selection of an appropriate formulation for a topical flea treatment for dogs and cats are good solubility of the compound, good adhesion to the skin, good spreading properties, good local and systemic tolerance, stability and compatibility within legal standards. The selected imidacloprid spot-on formulation, which meets all these requirements, contains 10 g a.i. in 100 ml nonaqueous solution. The product is dispensed in individual tubes (pipettes), available in different sizes. These are emptied completely after recommended usage thus reducing the possibility of ecological contamination (see also chapter 'Ecological Effects and Ecotoxicological Studies'). Advantage® is available in the following four sizes for dogs and two sizes for cats (in boxes with four or six applicator tubes):

- Advantage® 40 - for small dogs and puppies less than 4 kg (0.4 ml)
- Advantage® 100 - for dogs of 4-10 kg (1.0 ml)
- Advantage® 250 - for dogs of 10-25 kg (2.5 ml)
- Advantage® 400 - for dogs of 25-40 kg (4.0 ml), twice for dogs over 40 kg (8.0 ml)
- Advantage® 40 - for cats and kittens less than 4 kg (0.4 ml)
- Advantage® 80 - for cats of 4 kg and more (0.8 ml)

The compound is offered in a spot-on formulation which means it is applied externally by spotting the liquid onto the skin of the animal. The application is as follows: First, one applicator is removed from the package. Second, the applicator tube is held in an upright position, and the cap pulled off. Third, the cap is turned around and the other end of the cap placed on the tube. Last, the cap is twisted to break the seal and then removed from the tube. For easy application, the animal should be standing. For a proper application, to guarantee high efficacy with deposit into the lipid layer of the skin, the hair should be parted until the skin is visible, the tip of the tube should be placed onto the skin and the tube's contents should be applied directly onto the skin by squeezing the tube firmly twice. The site of application should be inaccessible to avoid licking and hence oral uptake of the product. This is especially important in small and young puppies. For example, an ideal inaccessible area for application is between the shoulder blades of dogs or on the back of the neck at the base of the skull of cats.

A potential limitation of any spot therapy is related to the body surface area of the animal. The larger the dog, the more limiting translocation of the ingredient can be. Applying multiple spots is a way to maximize the distribution of the a.i. and achieve as complete a coverage of the animal as possible (Arther and MacDonald 1997). Uniform distribution ensured by multiple spots also minimizes the possibility of product runoff (Arther and MacDonald 1997). In large dogs (>25 kg) the division of the dose on two sites is recommended for better efficacy. Young et al. (1996) reported 99% product efficacy for at least 28 days instead of 91.6% when the dose is not divided. Cunningham et al. (1997a) also recommended use of multiple sites in large dogs. The manner of application has been the object of a number of studies. A split application of 4 ml into four spots onto the skin of large dogs (>25 kg) confirmed high efficacy. For example 4 ml dose divided into four spots and 5 ml separated into two spots, both provided flea control rates above 99% through day 42 (Cunningham et al. 1997a). On day 21 the two-spot application was slightly less effective with 99.9% control, compared to the smaller amount of formulation split into four spots, which gave 100% efficacy on the same day (Cunningham et al. 1997a) (Fig. 17).

In trials with different volumes but a constant dose of 10 mg imidacloprid/kg body weight, the highest efficacy was rendered with the 10% formulation, comparing to higher concentrations e.g. a 20% formulation, reducing the volume by half or a 30% formulation, reducing the volume to one third. The larger the volume of the formulation, the better the spreading over the skin and the better the insecticidal efficacy (Fichtel 1998). It was concluded that a minimum volume of imidacloprid as a spot-on formulation is necessary to satisfy effectiveness. This explains the use of a 10% formulation even in larger dogs instead of using higher concentrated formulation with reduced volume (Fichtel 1998).

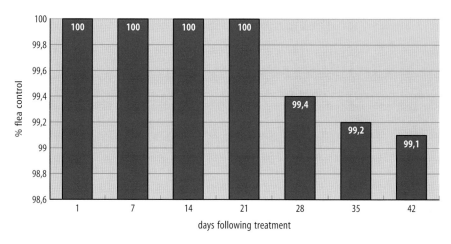

Fig. 17. Flea control persistency (in %) of imidacloprid 10% spot-on over a 42 days study period in large dogs (b.w. over 40 kg)

The correlation between ideal spreading of the compound and good efficacy again is the reason for division of the formulation in larger dogs instead of using higher concentrated formulation with reduced volume (Fichtel 1998), and this is confirmed by better efficacy results from divided applications by Fisher (1996). The dependency between efficacy and volume demonstrates the importance of the distribution of the formulation over the body surface with skin as carrier of the compound (Fichtel 1998) (Fig. 18).

After dermal application in dogs and cats, three media were discussed as potential carriers, which could result in exposure of the target species either orally of by simple contact. These were the fur or coat, the skin itself or the blood (Fichtel 1998). Studies finally identified the skin surface as the effective carrier of the compound after dermal application and the one producing the effect of imidacloprid, presented in the Advantage® formulation used for flea treatment (Fichtel 1998).

**Fur or coat**
Imidacloprid is spread through the coat of dogs and the fur of cats within twelve hours post application (p.a.) to cover the entire body and could be detected in studies of Fichtel (1998) in this medium over the whole 28 day period after application. In dogs the amount of imidacloprid in the coat from the trunk was insecticidal over the whole period of 28 days. But after 14 days p.a. this action was reduced. At the application site, the amount of compound showed good flea efficacy over the whole time (Fichtel 1998).

In Fichtel's (1998) study in cats insecticidal concentrations in the fur of the trunk could not be demonstrated, whereas the fur of the application site possessed very good pulicidal efficacy, although declined and was low 21 days p.a.

This pulicidal efficacy (Fichtel 1998) in the dog at least during the first half of the clinically effective period will be of importance also for a larvicidal effect in the environment after imidacloprid treatment of pets (see chapter 'Larvicidal Effect of Imidacloprid and Efficacy Enhancement (with PBO)'). The closer the contact of the flea with the application site, the higher the expected insecticidal activity will be. Considering the high mobility of fleas, contact with these areas of high concentration is expected at least for some time (Fichtel 1998).

Fichtel (1998) studied the spreading of the compound in detail. Maximum values of the mean a.i. concentrations twelve hours p.a. in the hair of the body were: 267 (shoulder) and 57 (flank) mg/kg for dogs and 27.25 (shoulder) and 27.13 (flank) mg/kg for cats. Thereafter the mean a.i. concentrations steadily decreased to a minimum at four weeks p.a. with 48 (shoulder) and 11 (flank) mg/kg on hair for dogs and 1.06 (shoulder) and 1.50 (flank) mg/kg for cats. Fichtel (1998) reported that average concentrations higher than ~100 mg/kg a.i. on hair showed good insecticidal effect in vitro. Average concentrations between 100 and 10 mg/kg a.i. on hair were significantly less effective, and concentrations below 10 mg/kg fur had almost no effect.

Other studies treating dogs with imidacloprid 10% spot-on at a dose of 10 mg/kg b.w. and collecting hair samples on various parts of the body six hours later for analysis of imidacloprid identified the concentrations listed in Fig. 19 and Fig. 20: The results of the study indicated that the compound was distributed quickly over the body surface allowing rapid contact and killing of adult fleas. Hair samples were collected in different groups at

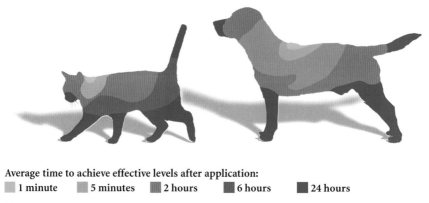

**Average time to achieve effective levels after application:**

■ 1 minute ■ 5 minutes ▥ 2 hours ■ 6 hours ■ 24 hours

Fig. 18. Distribution of imidacloprid after dermal spot-on application in cats and dogs (hours after treatment)

different time intervals and the contents in dog's coats were measured over a period of 42 days. It was shown that imidacloprid is distributed over the body of long- and short-hair coated dogs within one day and that levels were highest near the site of application and generally remained high for 21 days before declining between day 28 and 42 p.a..

Similar trials performed by Gyr and Hopkins (1995) detected imidacloprid in hair samples of treated dogs from different parts of the body at different times, up to 42 days after application. Levels were found over the body surface during the whole investigational period at all sites of the body from the neck to the caudal surface of the hind legs and on the tail. Concentration levels decreased with increasing distance from the application site, with mean concentrations highest on day 14 on the side of the chest (25.9 ppm). The highest concentration in a single sample was 212 ppm in one sample from the side of the neck on day 1. Good spreading properties on the body surface and persistence of the a.i. in the coat could be demonstrated (Gyr and Hopkins 1995).

In cats comparable trials with hair samples taken at twelve hours and 1, 7, 14, 21 and 28 days p.a. also revealed that imidacloprid is distributed over the body within twelve hours and remains on the skin for at least 28 days. Detailed mean concentrations on the hair are shown in Fig. 20.

### Skin

Another medium of importance in the mode of action of imidacloprid as flea adulticide is the skin and the skin surface.

Studies in cats using flea cages have shown that the uptake of imidacloprid out of the blood and/or the subcutaneous tissue causes small, reversible and non-lethal reactions in fleas (Bardt 1996). Similar results have been shown in dogs. Imidacloprid uptake out of the blood and/or the subcutis is therefore only minimally effective against fleas, whereas the uptake of the compound via the hair and the skin surface shows good insecticidal potency.

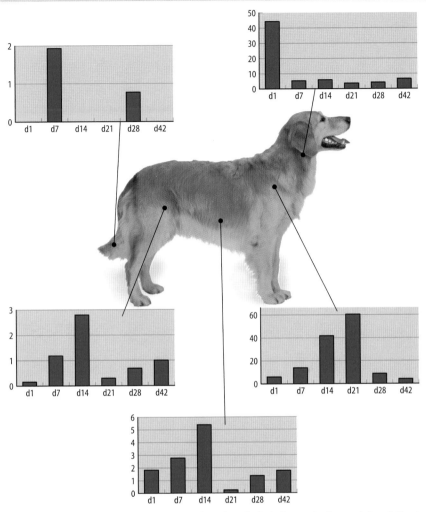

**Fig. 19.** Mean concentration of imidacloprid found on the hair (in ppm) of treated dogs following time p.t.

Very good efficacy is produced simply by the skin surface and is minimally increased by the compound localized in the hair (Fichtel 1998).

## Blood

Blood was examined as a potential carrier of the compound. It did not prove to play a role either in dogs or in cats. In dogs the levels were below the lowest pulicidal effective concentration of 0.1 mg/l serum over the whole period of eight weeks, most of the time even below 0.02 mg/l serum (Fichtel 1998). In cats the blood concentration of imidacloprid reached the minimally effective level of 0.1-0.2 mg/l serum only for a short period between ~2-48 hours p.a. (Fichtel 1998).

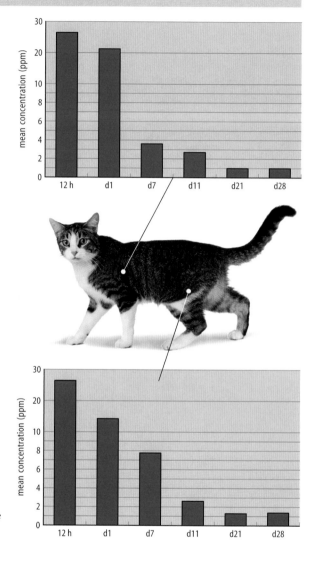

**Fig. 20.** Mean concentration of imidacloprid found on the hair (in ppm) of treated cats following time p.t.

These results and the spreading of the compound over the body via the skin demonstrate the skin as the important carrier for imidacloprid in flea treatment (Fichtel 1998). Similar and more detailed findings on the mode of spreading and action of imidacloprid as flea adulticide used as spot-on formulation for dermal application have been reported by Mehlhorn et al. (1999). They confirm the skin as main carrier and proved the compound to be localized in the water-resistant lipid layer of the skin surface, produced by sebaceous glands and spread over the body surface and onto the hair (Mehlhorn et al. 1999) (graphical description of the skin, Fig. 21). Spreading over the skin surface was reported by Everett et al. (2000a, 2000b) along with high flea control as early as six hours p.t.. Removal of the superficial fatty layer of the skin by repeated

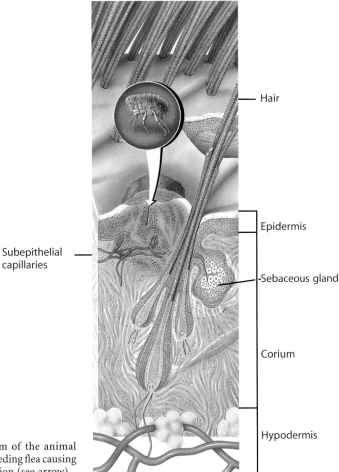

Hair

Epidermis

Subepithelial capillaries

Sebaceous gland

Corium

Hypodermis

**Fig. 21.** Dermogram of the animal skin with a blood feeding flea causing inflammatory reaction (see arrow)

swabbing with alcohol in treated dogs (clipped in the area of swabbing) resulted in feeding of the fleas with blood consumption comparable to the amount consumed by fleas in untreated controls (Mehlhorn 2000; Mehlhorn et al. 2001b). In contrast to this prepared area, in 'uncleaned', clipped areas on the same dogs fleas started feeding, but stopped within three to five minutes, starting to show tetanic and trembling movements and died within the first hour of exposure (Mehlhorn 2000; Mehlhorn et al. 2001b). Thus imidacloprid acts on adult fleas by contact (Mehlhorn 2000; Mehlhorn et al. 2001b). The compound is not taken up during sucking, but is absorbed via the fleas' thin intersegmental membranes (Mehlhorn et al. 1999).

On the basis of all the above mentioned results it can be concluded that imidacloprid, when used as flea adulticide by dermal application, is active by contact, in contrast to its classification into the category of systemic insecticides when formulated for agrochemical use (Fichtel 1998).

Besides the spread of the compound over the body surface, the initial rate of flea kill after treatment with imidacloprid is an important factor for optimal adulticidal efficacy. In studies with dogs, infested with 100 adult fleas and treated with imidacloprid, the compound demonstrated good flea control of approximately 89% efficacy against the existing flea populations at eight hours following product administration and 100% at twelve and 24 hours p.t. (Cruthers and Bock 1997). Other studies indicated a 98.5% efficacy six hours p.t. (Everett et al. 2000a, 2000b).

Once spread the rate of flea kill following weekly infestation is 93.3 to 100% of reinfesting fleas within two hours at days 7, 14 and 21 and 100% at eight hours at day 28 (Cruthers and Bock 1997) (Fig. 22).

In in vitro trials examining the effect of temperature on the toxicity of imidacloprid a.i. against adult cat fleas, the $LC_{50}$ ranged from 0.05 to 0.32 µg/cm², depending on the temperature (Richman et al. 1999). The compound was least toxic at 26°C and showed the highest toxicity of all the tested temperatures at 35°C (Richman et al. 1999) such high effectiveness can be expected at body temperature. The average fur temperature ±SD of two cats in studies by Richman et al. (1999) was 34.70 ± 0.27°C and temperatures on the cats' heads averaging 35°C, shoulders 34°C, abdomens 35.20°C, and tails 34.60°C. Average coat temperature of two dogs was 34.20 ± 0.58°C, head 35.70°C, shoulder 34.90°C, stomach 32.40°C, rump 33.40°C, and back leg 34.50°C (Richman et al. 1999). In this context it has to be mentioned that imidacloprid 10% spot-on is not inactivated by sunlight (Marsella 1999).

## Clinical Efficacy of Imidacloprid (Advantage®) in Dogs and Cats

Comparison of laboratory efficacy data is always difficult, because of differences in methodology, product concentrations and formulations and a lack of knowledge of insecticide susceptibility patterns of the flea strains used in the studies (Rust and Dryden

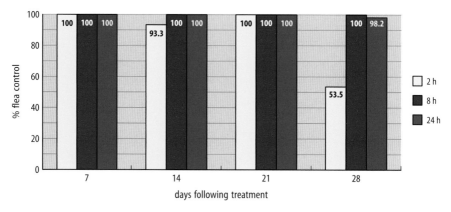

Fig. 22. Efficacy of imidacloprid on dogs at two, eight and 24 hours after weekly reinfestation with cat fleas

1997). Planning a clinical test of a compound is always a compromise between the standardization of the trial conditions on one side, considering the practicability and reasons of evaluation, and on the other side keeping conditions as close as possible to the clinical situation (Heeschen 1995). In this context, weekly reinfestations with 100 adult fleas are often used to simulate a continuous pressure of infestation from the natural surrounding onto an animal, repeatedly causing developing new generations of fleas. The efficacy of insecticide formulations under actual field conditions is generally difficult to assess because of the paucity of properly standardized published data (Rust and Dryden 1997). Formula to evaluate flea efficacy see Box 7.

---

**BOX 7.  Evaluation of efficacy**

**Formula according to Abbott's formula:**

**Efficacy (%) = 100 x $(m_c - m_T)/m_c$**

**Control group $(m_c)$: mean numbers of live fleas on the animal
Treatment group $(m_T)$: mean numbers of live fleas on the animal**

---

Numerous controlled efficacy studies were conducted according to standardized protocols at locations in the USA, Australia, and Europe to evaluate the initial and residual efficacy of Advantage® for control of both laboratory and field strains of *C. felis* on dogs. In each of the laboratory studies, the dogs were artificially infested with adult fleas one or more days prior to treatment and then periodically reinfested thereafter for the duration of the study. Dogs were randomized into groups to receive imidacloprid treatment or to serve as untreated or placebo controls. Treatments were accomplished by parting the hair over the shoulder blades and/or the iliac crest (rump) and applying the formulation directly onto the skin. In field studies usually client-owned dogs were treated in like manner after initial (baseline) flea counts were performed. Flea counts are usually performed at day one after treatment, followed by flea counts at special time intervals mostly weekly until study completion. In laboratory studies these were generally performed the day after reinfestation. The level of flea control is generally achieved by calculating the percentage reduction in flea numbers on treated animals compared to untreated or placebo-treated ones.

Two basic methods are reported for the technique of flea counts, manual counting and comb counting. Manual or so-called thumb counting is usually performed by the same technicians for all groups to eliminate bias. It represents a form of area count methodology (Dryden et al. 1994). Countings are usually limited to a special number of minutes per location and are conducted as described by Dryden et al. (1997): The hair is parted against the lay using both hands until the area is covered, for example in the locations dorsal midline, tail head, left lateral, right lateral, and inguinal region. In the second method, the so-called comb counting, fine tooth combs are used as tool for area and full body counts. Zakson et al. (1995) propose a standard of five minutes for comb counts. The accuracy of the two methods depends very much on the diligence and experience of the operators. Heckenberg et al. (1994) and Dryden et al. (1994) reported higher counts in dogs using comb counting. A difference of 73 fleas recovered by comb counting versus no fleas practicing thumb counting was even reported in one case by Heckenberg et al. (1994). Other authors state that comb counting is less dependent on operator experience and technique but that, with experienced operators, thumb counting gives accurate and repeatable results as well. Unpublished research data (Bayer Australia research data on file) over 25 years of testing the repeatability and comparative recovery of both methods (Hopkins et al. 1996a; Hopkins et al. 1997a) indicates that operator experience is the most important factor in sensitivity and repeatability (see Box 8).

---

**BOX 8. Important features in flea counting techniques**

➔ **specify personnel for the whole duration of a study**

➔ **describe and specify flea comb (e.g. extra-fine comb with 72 teeth, size 6.5 cm) when used**

➔ **specify area and order of body surface to be combed or covered by thumb counting (in case of area count)**

➔ **set time for combing (recommendation 5 min/ animal)**

➔ **comb or 'thumb' control animals first, use separate comb**

➔ **comb treatment groups from low to high dosages, use separate comb for each group**

---

### Laboratory Studies in Dogs

All in all, excellent efficacy has been reported with imidacloprid by numerous authors in laboratory studies, concluding that imidacloprid at a dosage of 10 mg/kg b.w. and monthly application is extremely effective for flea control in dogs.

Testing the efficacy of different formulations of imidacloprid as a spot-on preparation against fleas in dogs 7.5% and 10% solutions were completely effective in eliminating fleas for four and five weeks respectively, whereas imidacloprid 1% and 5% solutions controlled flea populations satisfactorily for one and three weeks respectively (Heeschen 1995). In a second study using a 10% imidacloprid formulation (applied dose 0.1 ml/kg), efficacy in eliminating fleas was recorded for four weeks (Heeschen 1995).

At different dosages there were differences in speed of flea kill based on flea counts 12 hours p.t. and also in duration of efficacy. The flea numbers were assessed 72 hours p.t./post reinfestation. The applications of 1 mg/kg b.w. and 5 mg/kg b.w. showed 100% effect for up to seven and 21 days respectively. With 7.5 and 10 mg/kg b.w. 100% efficacy was observed until day 28 p.t.. Treatment with 20 mg/kg b.w. in a 10% imidacloprid solution (i.e. 0.2 ml/kg b.w.) gave 100% efficacy for five weeks (Liebisch and Heeschen 1997). The knock-down effect of 100% was present for two weeks p.t. in treatments with 5 mg/kg, 7.5 mg/kg, and 10 mg/kg b.w. and stayed at >90% for four weeks in these test groups. In the 20 mg/kg b.w. group 100% knock-down effect remained for four weeks and >90% for the total of six weeks. In the treatment group with 1 mg/kg b.w. no distinct knock-down effect could be observed (Liebisch and Heeschen 1997).

The three main criteria of flea control using insecticidal therapy, the onset of efficacy, the duration of efficacy and the knock-down capability, have shown to be all positively related to dosage of imidacloprid in laboratory trials with dogs (Liebisch and Heeschen 1997). Summarizing the results of the above mentioned study, the 10% formulation administered at 10 mg/kg b.w. was considered due to the pharmacokinetic the most appropriate for further development (Liebisch and Heeschen 1997).

The superiority of 7.5 mg or 10.0 mg of imidacloprid/kg over 3.75 mg/kg for flea control was also observed by Arther et al. (1997b). In flea-infested and reinfested dogs treated with 3.75 mg imidacloprid/kg b.w. flea control ranged from 94.4 to 96.9% for days 14 to 28 and decreased to 91.6% by 34 days, whereas control with 7.5 and 10.0 mg/kg was 97.8 to 100% through day 28 for both. At day 34, the efficacy was 97.6 and 96.9% for 7.5 and 10.0 mg/kg b.w. respectively (Arther et al. 1997b). A summary of the study is that 9.1% imidacloprid solution applied as a low-volume spot treatment at dosages of 7.5 to 10 mg/kg kills fleas on dogs within 24 hours, with highly effective residual efficacy extending through 34 days (Arther et al. 1997b). Application at monthly intervals will break the flea life cycle because of its killing effect within 24 hours before egg production by adult fleas begins (Arther et al. 1997b).

In dose titration studies in Germany (Heeschen and Liebisch 1994), in two studies in the USA (Cunningham and Everett 1994; Cunningham 1995) and in Australia (Gyr et al. 1995) with either fixed dose volume but different concentrations of active (in the first two studies) or final formulation but different dose volumes (in the latter two studies), a clear dose dependent effect over the investigational period of up to twelve weeks could be demonstrated. Based on these results, a minimum dose of 10 mg/kg was considered to provide 100% knock-down efficacy within 24 hours p.t. and to protect against reinfestation for at least four weeks after a single treatment. This was investigated in further dose confirmation studies (see below) (Krieger 1996).

In the studies evaluating the initial and residual efficacy of imidacloprid for the control of fleas on dogs, animals were treated with placebo (the vehicle used in Advantage®) or 3.75, 7.5, receptively 10.0 mg/kg (in the formulation later on to be used for the product Advantage®). The study demonstrated that imidacloprid applied at the minimum target therapeutic dose of 10 mg/kg killed 99% of the fleas within one day of treatment and continued to provide 99-100% control of further flea infestations for at least four weeks (Cunningham and Everett 1994) (Fig. 23).

In two additional dose response studies the concentrations of imidacloprid were either 5.0, 7.5, 8.5 and 10.0 mg/kg b.w. or the plain vehicle of the formulation as negative control. Again both study results demonstrated that imidacloprid's minimum target therapeutic dose of 10.0 mg/kg provided 100% flea kill within one day after treatment. In the first of these trials, efficacy remained above 95% for three weeks and was 94% at

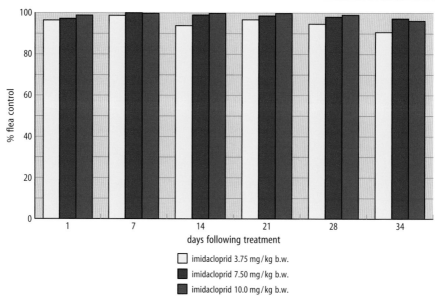

Fig. 23. Flea control (in %) achieved with three imidacloprid dosages in a dose-response study

day 28 and 93% at day 35. In the 8.5 mg/kg group results were similar, but efficacy declined more rapidly after day 35 compared to the 10.0 mg/kg group (Cunningham and Everett 1994) (Fig. 24). In further trials conducted in Australia, >95% control was provided through day 28 in the 10.0 mg/kg group.

Dose confirmation studies by Gyr (1995), Liebisch and Heeschen (1995) and Young (1995) all demonstrated a 100% knock-down efficiency within one day after treatment and a protective period against reinfestation for at least four and up to six weeks when imidacloprid spot-on was applied as a single spot on dogs weighing up to 25 kg and with split doses on dogs weighing more than 25 kg. Gyr (1995) treated dogs under 25 kg of weight in two controlled dose confirmation studies with a target dosage of 10 mg/kg and a minimum dose of 40 mg per animal. The results demonstrated 100% efficiency after two weeks, 97.5% after three, and 95.1% after four weeks, based on flea counts 24 hours after each reinfestation. No difference between different breeds and coat length could be observed in the trials. Liebisch and Heeschen (1995) investigated the efficacy in 20 Beagle dogs 24 hours after treatment and also 24 hours after reinfestation. Efficiency in their studies stayed at 100% during the first three weeks, declining to 98.4% on day 28, 95.7% on day 35 and 91.9% mean efficiency on day 42.

A final dose confirmation study with 10.0 mg/kg concentration of Advantage® indicated again that the minimum target therapeutic dose of 10.0 mg/kg killed 100% of infesting fleas within one day of application, and showed a residual efficacy remaining for >96% for at least four weeks (Fig. 25). In Hopkins et al. (1996a) the therapeutic

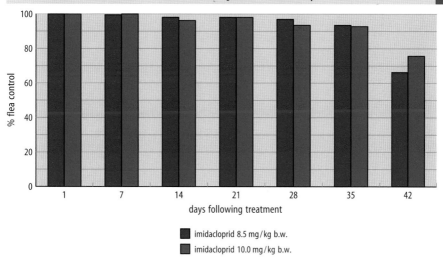

Fig. 24. Flea control (in %) achieved with two imidacloprid dosages in a dose-response study

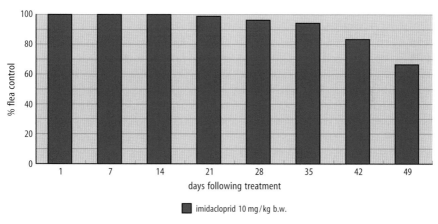

Fig. 25. Flea control (in %) achieved with imidacloprid 10 mg/kg b.w. as confirmed by a dose-confirmation study

dosage of 10 mg imidacloprid/kg b.w. also provided 100% removal of flea infestations from dogs at 24 hours after treatment and more than 95% control of reinfestation, at 24 hours after reinfestation, for four weeks.

Evaluation studies testing the initial and residual flea control efficacy of 9.1% imidacloprid solution in laboratory trials with dogs revealed 99-100% of flea control on day 1 p.t. in groups being treated at a single or at two sites. The single application site group had ≥91.6% efficacy on days 7-28 declining to 74.5% by day 42, whereas the two-spots group showed an efficacy of >99% through day 28, declining to 96.3% and 92.4% on day 35 and 42 p.t., respectively (Young et al. 1996). This emphasized the importance of a division of the application sites and generally confirmed the efficacy of imidacloprid against flea infestation (here in large dogs) over the period of (at least) one month.

Other trials conducted by Hopkins et al. (1997a) with 10% w/v topical formulation of imidacloprid showed 100% efficacy at day 1 and week 1, compared to an untreated group of dogs. Further 99% efficacy at two weeks, 97.5% at three weeks, and 95.1% at four weeks could be recorded. Thus the treatment was highly effective in removing existing flea infestations within 24 hours and preventing reinfestation for up to four weeks (Hopkins et al. 1997a) (Fig. 26). The condition for successful flea treatment is the removal of fleas in less than 36 hours to break the flea life cycle (Hopkins et al. 1997a), because female fleas need 36 to 48 hours to feed and mature before they are able to lay fertile eggs (Osbrink and Rust 1984).

Artificial reinfestations with adult fleas in the above studies (Hopkins et al. 1997a) were made away from the application site but also on the back of the animal. To exclude any bias effect for the treatment by flea infestations in proximity to the site of application, subsequent work with artificial infestations applied to the underline of the animals was performed with same results, i.e. high efficiency of imidacloprid unrelated to the site of flea infestation (Hopkins et al. 1997a).

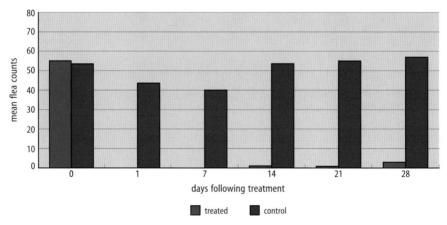

**Fig. 26.** Flea counts on imidacloprid-treated dogs in comparison to untreated control animals

A second study by Hopkins (1997a) was conducted to test the efficacy of imidacloprid (10% w/v) topical formulation in infested dogs which were reinfested with 150 fleas weekly for seven weeks and examined for fleas 30 hours after every artificial reinfestation. Until day 14 p.t. 100% efficacy was observed. Day 21 revealed 99.1%, day 28 99.6%, day 35 99.3%, day 42 98.1%, and finally day 49 88.1% efficacy (Hopkins 1997a). Nearly 100% flea control for at least four weeks by dermal application of 10% imidacloprid spot-on formulation (10 mg/kg) in adult dogs of various breeds compared to controls was also observed by Werner et al. (1995b). In detail, until day 15 (after weekly infestation with 100 fleas) the mean flea counts remained at zero. On days 22 and 28 a small number of fleas could be recovered (Fig. 27).

## Field Studies in Dogs

The difficulty of transmitting laboratory results into field conditions has been mentioned already (see chapter 'Clinical Efficacy of Imidacloprid (Advantage®) in Dogs and Cats'). The field performance of imidacloprid has been tested in a number of studies, encompassing the range of possible flea strains, environmental conditions etc..

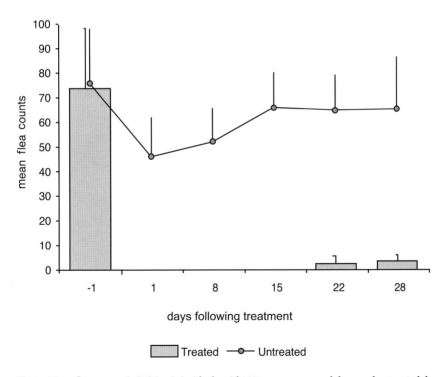

**Fig. 27.** Mean flea counts (+SD) in six imidacloprid 10% spot-on-treated dogs and untreated dogs one day after weekly reinfestation with 100 fleas

In eleven veterinary practices in France 70 dogs of both sexes, all ages, and various breeds and coat lengths were chosen on the basis of flea infestation and signs of flea allergy (Griffin et al. 1997). The dogs were treated with imidacloprid 10% w/v spot-on formulation. Dogs between 1.0-4.0 kg b.w. received 0.4 ml, between 0.4-10.0 kg b.w. 1.0 ml, between 10.0-25.0 kg b.w. received 2.5 ml, and between 25-50 kg b.w. 5.0 ml divided into two spots. Weekly mean efficacy values for imidacloprid spot-on under constant flea exposure were 93.6% on day 1 p.t., 96.1% on day 7 p.t., 94.9% on day 14 p.t., 93.55% on day 28 p.t., and 90.0% on day 35 p.t. (Griffin et al. 1997). Besides a reduction in flea numbers both prevalence and severity of pruritus and subsequent frequency of scratching decreased following treatment. A reduction in the number of dogs exhibiting alopecia as well as a decrease in degree and frequency of FAD lesions were observed (Griffin et al. 1997). Thus a high efficacy over a period of 28 to 35 days following a single treatment with imidacloprid 10% spot-on formulation (10 mg/kg b.w.) could be observed in this field study in France, with decreasing effects of flea infestations on the animal within one week of treatment and a marked decrease in pruritus, alopecia, and FAD by 28 days p.t.. Based on these field results imidacloprid at 10 mg/kg b.w. administered monthly can be expected to give excellent results in cases of FAD (Griffin et al. 1997) (Fig. 28). A second clinical trial including 140 naturally flea-infested dogs

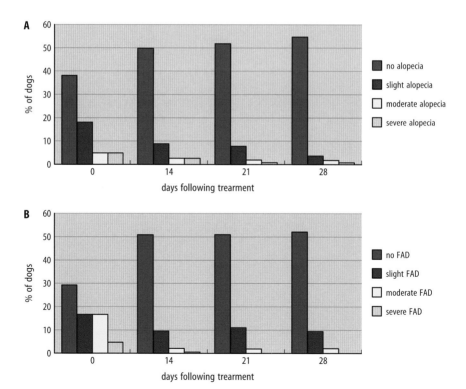

**Fig. 28.** Clinical reparation of alopecia (**A**) and decrease of flea allergy dermatitis symptoms (**B**) after a single topical treatment with imidacloprid in flea-infested dogs

demonstrated mean reductions in flea counts of 93.6% on day 1 p.t., 96.1% on day 14 p.t., and 90.1% on day 35 p.t. (Bayer Corp. USA research data on file). In this study 71% of the dogs were exhibiting pruritus, alopecia, and other signs of FAD on the first day of treatment which dropped to 19% by day 28 (Bayer Corp. USA research data on file).

Field studies to measure the efficacy of flea control over three summer flea seasons were carried out in Australia. The results were summarized by Hopkins (1997c) (see also Table 21): Twenty two sites with 59 dogs (52 pure dog households and seven households with dogs and cats) and a history of treatment failure of flea infestations as well as continuing reinfestation from surroundings were included in the study. Ten percent imidacloprid w/v topical formulation at either the minimum proposed therapeutic dosage of 10 mg active/kg b.w or the proposed recommendations for veterinary use (here 0.4 ml up to 4 kg, 1 ml up to 10 kg, 2.5 ml for 10-25 kg, and two doses of 2.5 ml for >25 kg b.w.) at monthly intervals were tested. Flea control was defined as reduction of flea numbers on the dogs to zero for at least two months after the last treatment. Fifty-two dogs (i.e. 88%) showed no fleas one day after treatment, three had low, two moderate numbers of fleas, and two showed heavy or very heavy reinfestation. At day 7 p.t., 40 dogs had no fleas, 15 low numbers, two moderate levels and two high. By four weeks, 43 dogs had no fleas, twelve had low numbers, two moderate and two still high levels of reinfestation. By seven weeks only low to medium levels of infestation were found on two out of the initial 59 dogs, which were on an extensive property including a dam for daily swimming. Three treatments were necessary to resolve their problem, and by nine weeks elimination of this reinfestation was also achieved.

At twelve sites, i.e. in 55%, a single treatment was sufficient to completely resolve the flea control problems. In eight cases, i.e. in 36%, two treatments were needed. Three treatments were necessary at two sites where dogs were swimming on a regular basis (Hopkins 1997c) (Fig. 29). Abatement of clinical signs of FAD was seen in all cases within seven days (Hopkins 1997c). Furthermore regression of skin pathology and major hair regrowth, in cases with alopecia, were recorded within four to six weeks after the first treatment (Hopkins 1997c). Imidacloprid proved it could resolve and eliminate long standing flea problems (on 59 dogs at 22 sites) without any other chemical or management method to control fleas (Hopkins 1997c). Nevertheless further treatments on a monthly basis are recommended especially at sites with possible external sources of reinfestation and animals with FAD (Hopkins 1997c).

Testing the efficacy of imidacloprid in the removal of fleas and the resolution of FAD 1939 dogs (604 from single-animal households and 1335 from multiple-animal households) were treated in collaboration with 423 veterinary clinics located across Italy (during spring/summer 1997) (Genchi et al. 2000). Before treatment animals were clinically examined, the flea burden was assessed by flea thumb count, and dermatitis lesions of FAD were ranked according to severity of typical clinical signs. Dogs were treated with 10% w/v imidacloprid in a single spot on the back line at the recommended dosage (minimum dosage: 10 mg/kg b.w.), and no concomitant treatment (also not in the environment) was given. Flea numbers dropped significantly after treatment in

Table 21. Control of flea infestation in dogs in an Australian field study: regression of dogs with a flea burden throughout the study

|  | No fleas | Low levels | Moderate levels | High to very high |
|---|---|---|---|---|
| day 1 p.t. | 52 | 3 | 2 | 2 |
| day 7 p.t. | 40 | 15 | 2 | 2 |
| 4 weeks p.t. | 43 | 12 | 2 | 2 |
| 7 weeks p.t. | 57 | 2 (low to medium) | - | - |
| 9 weeks p.t. | 59 | - | - | - |

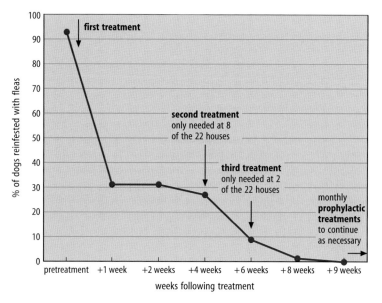

**Fig. 29.** Decreasing incidence of reinfestation of dogs with fleas from the surroundings after one, two, or three applications of imidacloprid 10% spot-on, different sites taken into account

dogs from both single- and multiple-animal households. Twenty-one days after treatment, the percentages of dogs free of fleas were 98% and 92%, respectively. At the same time, the mean efficacies (defined as the percent variation of the average of the sum of given scores at different times of the study: Wilcoxon Matched-Pairs Signed-Ranks Test) were 96% in dogs from single- and 90% from multiple-animal households (Table 22 and Table 23). Concerning the resolution of FAD signs, a total of 688 dogs (232 (38%) from single- and 456 (34%) from multiple-animal households) showed clinical signs of FAD prior to treatment, decreasing from 38% to 16% by day 14, 7% by day 21, and 6% by day 28 in single-animal households. In multiple-animal

Table 22. Control of FAD in dogs in an Italian field study: dogs without fleas throughout the 28 days study period

| | Days after treatment | | | | |
|---|---|---|---|---|---|
| | 0 | 1 | 14 | 21 | 28 |
| Single-animal households (No. of dogs) | 0 (604) | 60% (365) | 95% (574) | 98% (593) | 93% (561) |
| Multiple-animal households (No. of dogs) | 0 (1335) | 51% (678) | 88% (1173) | 92% (1228) | 83% (1105) |

Table 23. Control of FAD in dogs in an Italian field study: flea efficacy throughout the 28 days study period

| | Days after treatment | | | | |
|---|---|---|---|---|---|
| | 0 | 1 | 14 | 21 | 28 |
| Single-animal households (604 dogs) | 0 | 73% | 97% | 99% | 96% |
| Multiple-animal households (1335 dogs) | 0 | 65% | 93% | 96% | 90% |

households, FAD signs decreased from 34% to 17% by day 14, 10% by day 21, and 8% by day 28, thus verifying a rapid adulticidal and high residual activity, lasting at least four weeks. There was effective control of parasites with rapid improvement of allergy signs until almost complete remission up to 28 days following the first product application (Genchi et al. 2000).

In two German field studies 380 infested dogs exposed to possible reinfestation were treated with Advantage® (according to the manufacturer's recommendation). Acute and residual activity of the compound was tested by evaluating flea numbers before, one week after for acute or three to four weeks after treatment for residual activity, calculated on different sets of animals. The results were expressed as the percentage of animals with no fleas with the criterion of total elimination which was a more stringent one (compared to traditional methods of calculating percent reduction of flea burdens). Out of 59 cases 81% of the dogs had no fleas one week after treatment, demonstrating acute activity, and of 321 cases 92% had no fleas three to four weeks after treatment, representative of residual activity. All remaining animals in both groups showed a markedly reduced degree of infestation, with the majority of them possessing fewer than three fleas. All in all, the excellent efficacy of imidacloprid 10% spot-on under field conditions could be demonstrated (Ewald-Hamm et al. 1997).

Besides the criteria of percent reduction of flea burdens and of a number of cases of total elimination of fleas, a study at the University of California at Davis concentrated solely on the resolution of FAD symptoms. Fifteen dogs were treated with imidacloprid every three weeks for three months, resulting in the resolution of all clinical signs after three months in 40% of all dogs. Sixty percent presented moderate improvement, and all dogs displayed greater than 50% improvement in pruritus (Gortel 1997).

A similar object was examined by Hopkins (1998) in a study with nine dogs showing classic signs of FAD without secondary bacterial infection. These were treated solely with imidacloprid 10% w/v according to label recommendations. Six of the test animals were housed in covered pens with outside runs after treatment. These pens were hosed out daily, while the three remaining were left in their households. Two of these household dogs were retreated after one month. All penned dogs were free of fleas within 24 hours. The household dogs showed reduced numbers of reinfesting fleas up to 14-33 days p.t., after which no more fleas were seen. The incidence of scratching and biting the skin fell markedly within 24 hours in all dogs, being virtually zero by day 25 p.t.. By two weeks, most areas of exudative dermatitis were dry and hair regrowth was evident. By three weeks, no exudative lesions were found in any of the dogs, and by four weeks, scratch reflexes were absent. By 44-47 days, hair regrowth had covered most of the hairless areas. So it was concluded that resolution of uncomplicated FAD can be achieved using solely imidacloprid by a single dose and removal from sources of flea reinfestation, or in the field, by one or more applications within a flea control program (Hopkins 1998) (Fig. 30).

In more than 300 naturally infested dogs of various breeds, in pregnant and lactating bitches, in puppies and old dogs, alone or in combination with commonly used veterinary products, the efficiency of imidacloprid spot-on 10% was also tested under field conditions by Schein (1996), Jacobs (1995), and Hopkins (1995). The treatment on the basis of recommended dosage (at that time 0.4 ml for <4 kg, 1 ml for 4-10 kg, 2.5 ml for

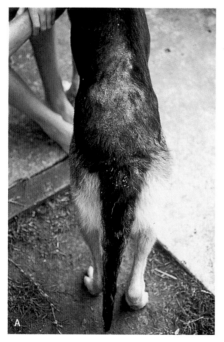

Fig. 30. A-F. Clinical signs of FAD in dogs and resolution after imidacloprid 10% spot-on treatment

10-25, 5 ml for >25 kg b.w. (the last is no longer the recommended dose, the current dosage is 4 ml for dogs >25 kg b.w.)) provided a high level of acute and residual efficacy for four to six weeks against cat and dog fleas (Krieger 1996).

Fig. 30. C-D.

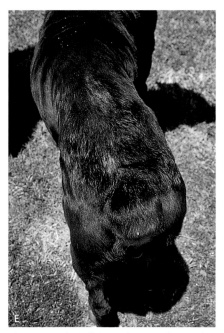 

Fig. 30. A-F. Clinical signs of FAD in dogs and resolution after imidacloprid 10% spot-on treatment

## Laboratory Studies in Cats

The general considerations of laboratory studies versus field trials are the same for cats as they are for dogs (see above).

A series of controlled efficacy studies was conducted under standardized protocol design at locations in the USA, Australia, and Europe to evaluate the initial and residual efficacy of Advantage® against various flea strains (*C. felis*) also on cats. In each of these studies the cats were experimentally infested with fleas one or more days prior to treatment and reinfested with additional fleas at definite intervals after treatment. The cats were randomized into groups to receive either a specified treatment dosage or to serve as untreated controls. The treatments were administered on day 0. The product was applied by parting the hair and applying the formulation directly onto the skin at the back of the neck. Live flea counts were then conducted on each cat for a certain amount of days usually the day after treatment and following each infestation.

The ideal concentration of imidacloprid for flea control has been studied particularly in dogs. Clinical studies in cats mainly concentrate on the efficacy of this dosage which is also commercially available. In dose confirmation studies with imidacloprid at 10.0 mg/kg b.w. >99% killing of infesting fleas within one day of treatment and a residual flea control which remained above 96% for three weeks, declining to 86.1% at four weeks, could be demonstrated (Fig. 31, Fig. 32).

Werner et al. (1995b) reported of the clinical efficacy of 10% imidacloprid formulation applied dermally to cats which were reinfested weekly. Nearly 100% flea control could

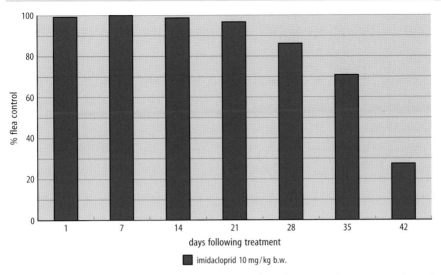

Fig. 31. Flea control (in %) achieved with a single imidacloprid treatment in cats over a six week period with weekly reinfestations

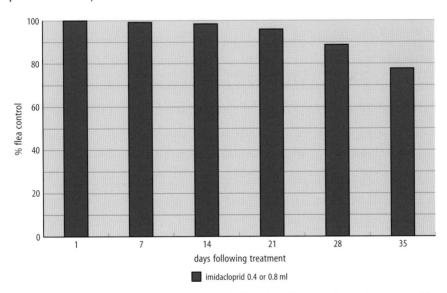

Fig. 32. Flea control (in %) achieved with two imidacloprid application volumes in cats: 0.4 ml for cats up to 4 kg and 0.8 ml for cats over 4 kg b.w.

be observed for at least four weeks in adult cats. Very small numbers of fleas were found on day 14, 21, and 28 p.t. hardly with any increase (Fig. 33). Further trials with 20 cats artificially infested with fleas, reinfested weekly for four weeks and half of them treated with 10% w/v topical formulation of imidacloprid in a single spot on the backline revealed a mean percentage reduction in flea numbers compared to the untreated group

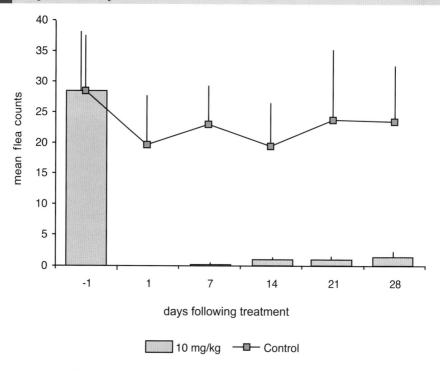

**Fig. 33.** Mean flea counts (+ SD) in ten treated (10 mg/kg b.w. imidacloprid 10 % spot-on) and ten untreated cats 24 hours after weekly infestation with 100 fleas

of 99.5% at day 1, 99.1% at week 1, 96.9% at week 2, 96.2% at week 3, and 95.7% at week 4 (Hopkins 1997a; Hopkins et al. 1997a). The treatment effect within 24 hours was as in dogs highly significant at all times in the cat trial (Hopkins et al. 1997a) as in the dog trial.

Considerations about the efficacy within 24 hours, to check for a break in the flea life cycle as well as the effects of reinfestation of animals away from the application site have already been outlined previously (see chapter 'Laboratory Studies in Dogs').

Young et al. (1996) tested the potential of imidacloprid 10% spot-on applied according to the label recommendations at the dorsal mid-cervical spine region versus blank vehicle in cats infested and weekly reinfested with 100 unfed adult cat fleas. They observed 100% flea control on day 1 p.t. and >96% control for test days 7-21. In their trials, the efficacy declined to 89.1% and 75.4% on test days 28 and 35, respectively (Young et al. 1996).

In trials of Jacobs et al. (1997a) artificially infested cats were treated with imidacloprid 10% (w/v) spot-on formulation applied at a dose rate of 10 mg/kg on the dorsal midline just behind the base of the skull. Significant differences between treated and control group members were detected with an overall efficacy of 99.8% at 24 hours p.t. and 100% at 48 hours p.t.. Twenty-four hours after the first reinfestation on day 7 no live fleas were found. Efficacy values in excess of 95% persisted further for between three to four weeks, established one day after each challenge and for between four to five weeks even when fleas were counted 48 hours after reinfestation (Jacobs et al. 1997a) (Fig. 34).

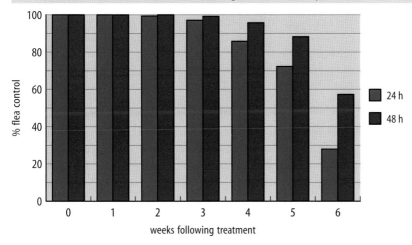

**Fig. 34.** Duration of efficacy against *C. felis* of imidacloprid applied topically to cats at 10 mg/kg on day 0: percentage efficacy values based on flea-counts made 24 and 48 hours after treatment and each subsequent infestation

Imidacloprid proved to possess considerable potency against adult fleas on cats and to retain a high level of activity for four to five weeks, so demonstrating a potential as an adulticide for both immediate relief from fleas and use in longer term flea control strategies (Jacobs et al. 1997a). The duration of protection of imidacloprid 10% spot-on against reinfestation with cat fleas after a single treatment is shown in Table 24.

**Table 24.** Geometric mean flea counts of two groups of cats: one untreated, the other treated with imidacloprid 10% spot-on at 10 mg/kg b.w. on day 0

| Weeks after treatment | Mean flea counts | | | | | |
|---|---|---|---|---|---|---|
| | After 1 day | | | After 2 days | | |
| | Control | Treated | Percent reduction | Control | Treated | Percent reduction |
| 0 | 36.7 | 0.01 | 99.8 | 31.7 | 0.0 | 100.0 |
| 1 | 34.1 | 0.0 | 100.0 | - | - | - |
| 2 | 31.0 | 0.2 | 99.3 | 27.3 | 0.0 | 100.0 |
| 3 | 32.1 | 1.0 | 97.0 | 28.9 | 0.1 | 99.7 |
| 4 | 36.2 | 5.0 | 86.1 | 27.8 | 0.9 | 96.7 |
| 5 | 33.8 | 9.5 | 71.9 | 26.8 | 3.1 | 88.3 |
| 6 | 28.1 | 20.1 | 28.4 | 24.8 | 10.6 | 57.3 |

A second study concentrated on the longer term prophylactic protection using the same formulation with monthly treatments in a controlled simulated home environment. As part of a larger experiment (Jacobs et al. 1997b) cats were artificially infested with fleas in the first phase and placed in a so-called controlled simulated home environment. This was carpeted pens in which the flea life cycle was initiated by placing a number of eggs, larvae and pupae on each carpet, thus allowing the flea population to propagate without any intervention except for the scheduled treatments of the cats described below and welfare combings in cases of excessive grooming behavior. In a second phase (starting on day 56), five new fleas were placed weekly on each cat with a flea burden below 30, to mimic roaming of the cats and acquisition of infections from outside the home. The cats were treated with imidacloprid as a 10% spot-on formulation at intervals of 28 days at the recommended commercial dose rate. Flea counts were performed at 14 day intervals and fleas replaced after counting up to a maximum of 30. Control cats showed an explosion of the flea population around day 42 after an initial drop of on-animal flea numbers, probably due to grooming activity (Dryden 1989b). This made welfare combings necessary. No fleas were detected at any time on the imidacloprid-treated cats and no welfare combings were required (Jacobs et al. 1997b). The first phase of the study (up to day 56) was designed as a closed epidemiological system representing cats confined to a hypothetical household, while phase 2 added the dimension of cats roaming and acquiring small numbers of fleas from outside the home. Non-detectable flea levels in the treated pens over the whole time of the study could be explained by residual insecticidal properties of imidacloprid killing all emergent fleas moving from the carpet onto the treated cats and/or adulticidal action of the compound causing too low egg production for the maintenance of the flea life cycle in the environment and also traces of compounds being transferred from the animals' coat to the carpet exerting larvicidal activity (Hopkins et al. 1996b). Regardless which of these three actions is responsible, the study proved that monthly imidacloprid treatments provide a very high level of protection, here over 16 weeks, despite severe challenge conditions (Jacobs 1997) (Table 25).

Table 25. Individual and geometric mean flea counts of two groups of cats: one untreated, the other treated with a control program using imidacloprid 10% spot-on at 28 day intervals

| Group | Cat no. | Flea counts on day | | | | | | | |
|---|---|---|---|---|---|---|---|---|---|
| | | 14 | 28 | 42 | 56 | 70 | 84 | 98 | 112 |
| Control | 1 | 8 | 6 | 94 | 73 | 46 | 46 | - | - |
| | 2 | 5 | 6 | 43 | 60 | 67 | 39 | 55 | 87 |
| | 3 | 2 | 10 | 50 | 60 | 38 | 37 | 58 | 104 |
| | 4 | 14 | 8 | 117 | 142 | 50 | 70 | 74 | 95 |
| | Mean | 6.0 | 7.4 | 69.8 | 78.2 | 49.2 | 46.5 | 61.8 | 95.1 |
| Imidacloprid | 5 | 0 | 0 | 0 | 0 | 0 | 0 | - | - |
| | 6 | 0 | 0 | 0 | 0 | 0 | 0 | 0 | 0 |
| | 7 | 0 | 0 | 0 | 0 | 0 | 0 | 0 | 0 |
| | 8 | 0 | 0 | 0 | 0 | 0 | 0 | 0 | 0 |
| | Mean | 0 | 0 | 0 | 0 | 0 | 0 | 0 | 0 |

- = no counts as these cats removed from trial (due to injury of one cat)

Concerning the site of product application only 60% protection was achieved at day 28 when the product had been applied between the shoulder blades of cats. This is likely to be a result of grooming behavior in cats, thus emphasizing the importance of correct application for efficacy in that animal species (Marsella 1999) namely on the back of the neck at the base of the skull. Generally a minimum therapeutic dosage of 10 mg/kg b.w. of imidacloprid in a topical formulation results in complete removal of existing flea populations on cats and provides a protective period from reinfestation for up to four weeks (Hopkins et al. 1997a) (Fig. 35).

In cats 100% reduction in the adult flea population was seen within 24 hours after application on the neck. After four weeks 89-96% reduction was observed (Hopkins et al. 1997a; Jacobs et al. 1997a; Young et al. 1996).

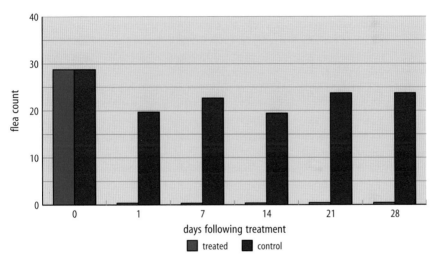

Fig. 35. Flea counts on imidacloprid-treated cats in comparison to untreated control animals

## Field Studies in Cats

The efficacy of imidacloprid in a solely European field study in cats was reported by Ewald-Hamm et al. (1997). Two hundred and twenty-five flea-infested cats which were exposed to possible reinfestation were included in these trials. A proportion were living in multi-pet environments. Flea numbers were evaluated by means of combing before treatment, one week after treatment for the assessment of acute activity, or three to four weeks after treatment for residual activity. These two criteria (acute and residual) were calculated from different sets of animals. After treatment with imidacloprid 10% spot-on, the percentage of animals with no fleas, which is more stringent than calculating percent reduction of flea burdens, was 82% (out of 57 cases) testing acute activity and 83% (out of 168 cases) testing residual activity, both confirming high efficacy of this formulation in cats also under field conditions (Ewald-Hamm et al. 1997).

In a second Italian study during spring/summer 1997 with 423 veterinary clinics including 1333 cats (291 from single-animal households and 1042 from multiple-animal households), imidacloprid 10% w/v topical formulation was tested at the recommended dosage (minimum dosage: 10 mg/kg) with regard to its efficacy in flea control and resolution of FAD signs (Genchi et al. 2000). Before treatment, animals were clinically examined, the flea burden was assessed by flea thumb count, and dermatitis lesions of FAD were ranked according to severity of typical clinical signs. No concomitant treatment (to cats or the environment) was given. Twenty-one days after treatment, the percentages of cats free of fleas were 98% and 91% from single- and multiple-animal households, respectively. At the end of the study (day 28) the percentages were 93% and 85%, respectively. At the same time the mean efficacies (defined as the percent variation of the average of the sum of given scores at different times of the study: Wilcoxon Matched-Pairs Signed-Ranks Test) were 95% in cats from single- and 90% from multiple-animal households (Table 26, Table 27). Concerning the resolution of FAD signs, a total of 323 cats (76 (26%) from single- and 247 (24%) from multiple-animal households) showed clinical signs of FAD prior to treatment, decreasing from 27% to 12% on day 14, 8% on day 21, and 5% on day 28 in single-animal households. In multiple-animal households FAD signs decreased from 24% to 11% on day 14, 6% on day 21, and 5% on day 28. Also in cats this verified a rapid adulticidal activity and high residual activity, lasting at least four weeks, with rapid improvement of allergy signs until almost complete remission at 28 days following the first product application under field conditions (Genchi et al. 2000) (Fig. 36).

In a field study in France, 60 flea-infested cats were treated with imidacloprid 10% spot-on while 55 other flea-infested cats served as controls and were treated with another adulticide (fipronil). Both groups represented a cross-section of the feline population

Table 26. Control of FAD in cats in an Italian field study: cats without fleas throughout the 28 days study period

| | Days after treatment | | | | |
| --- | --- | --- | --- | --- | --- |
| | 0 | 1 | 14 | 21 | 28 |
| Single-animal households | 0 | 57% | 96% | 98% | 93% |
| (No. of cats) | (291) | (167) | (279) | (285) | (272) |
| Multiple-animal households | 0 | 54% | 89% | 91% | 85% |
| (No. of cats) | (1042) | (563) | (931) | (952) | (883) |

Table 27. Control of FAD in cats in an Italian field study: flea efficacy throughout the 28 days study period

| | Days after treatment | | | | |
| --- | --- | --- | --- | --- | --- |
| | 0 | 1 | 14 | 21 | 28 |
| Single-animal households | 0 | 69% | 97% | 99% | 95% |
| (291 cats) | | | | | |
| Multiple-animal households | 0 | 64% | 93% | 95% | 90% |
| (1042 cats) | | | | | |

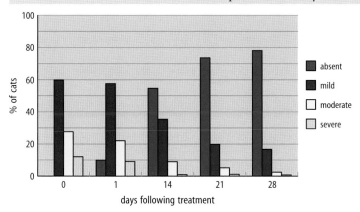

Fig. 36. Flea allergy dermatitis in 323 cats from multiple-animal households

and the individual signs of FAD were clinically scored according to a scale. Only one treatment was administered and animals were reexamined on days 1, 14, 21, 28, and 35 after treatment. Further subjective assessments of clinical signs were also made on each occasion. Concomitant treatment for skin and other diseases were allowed due to welfare reasons and owner compliance. These treatments were also likely to improve the clinical signs of FAD. The major clinical sign of FAD, pruritus, responded well to treatment with imidacloprid 10% spot-on while the effect on alopecia was less marked due to an insufficient time interval (28 days) to allow hair regrowth (Mason 1998). However, a general beneficial effect on the clinical signs of flea related disease in the majority of treated cats with imidacloprid 10% spot-on was evident (and comparable to the one induced by fipronil) (Mason 1998).

## Comparative Studies
## (Between Imidacloprid and Other Compounds for Flea Control)

Imidacloprid for flea control has been compared in its efficacy at various concentrations with other compounds in dogs and cats.

### Laboratory Studies

Comparing the products Advantage®, Frontline® Top Spot™, and Frontline® Spray (fipronil, Merial) in terms of convenience, safety, and duration of efficacy, the first two are both available in premeasured tubes designed for specific body weights of dogs and cats (to ensure accurate dosing). Advantage® is labeled for flea control, while Frontline® Top Spot™ has label claims for fleas and ticks. Their application to the pets' skin is similar, whereas their application period can be different: Advantage® is recommended monthly, while Frontline® Top Spot™ is recommended at one to three months intervals, depending on owners detecting parasites and infestation levels. Frontline® Spray is labeled for flea and tick control on cats and dogs and is applied every one to three months. The quantity applied is determined by body weight, hair coat type, and pest intensity. The spray requires a more rigorous application, resulting in a greater chance for animal intolerance, compared to the two spot-on formulations. For Frontline® Spray the user must wear household latex gloves during application (product leaflet Frontline® Spray).

Marchiondo et al. (1999) evaluated the speed of kill of imidacloprid (Advantage®) and two fipronil formulations (Frontline® Spray and Frontline® Top Spot™) on 32 artificially flea-infested dogs observing similar and excellent efficacy in all three products at 24 hours after each reinfestation and high levels of efficacy by all three products as assessed eight hours after challenge. They observed significantly higher efficacy using Frontline® Top Spot™ compared with Advantage® only at the flea counts eight hours after reinfestations on days 21, 28, and 43 (Marchiondo et al. 1999).

Imidacloprid has further been compared in laboratory studies in its efficacy and its speed of killing with the adulticide fipronil in fleas and ticks (Cruthers et al. 1999). Sixty artificially flea (*C. felis*)- and brown dog tick (*Rhipicephalus sanguineus*)-infested dogs were treated according to label instructions either with Advantage® (imidacloprid) or Frontline® Top Spot™ (fipronil), or left as untreated controls. Live fleas were recovered from each treated dog examined at six and twelve hours p.t. with either of the compounds, although a decline in the number was seen at each of these evaluations. At 18, 24, or 48 hours p.t. no live fleas were recovered from any treated dog. Tick counts on fipronil-treated dogs were lower at each post-treatment examination than on imidacloprid-treated dogs, and no live ticks were found on the former ones after the 24-hour evaluation. On the imidacloprid-treated dogs live ticks were present at all evaluations. Both compounds achieved complete elimination of an existing flea infestation between twelve and 18 hours p.t.. Ticks were completely eliminated from fipronil-treated dogs between 24 and 48 hours p.t. (Cruthers et al. 1999).

The efficacy of imidacloprid and fipronil at recommended dose rates (as positive control) was tested on artificially infested dogs and cats (Hopkins 1997a). The latter revealed only 84.6% instead of 100% in the imidacloprid-treated group 24 hours after treatment. Later on the two compounds were comparable, possessing 100% at seven and 14 days p.t.. Finally fipronil showed 100% efficacy at 21 days while imidacloprid possessed 99.1% efficacy at that time (Hopkins 1997a). The percentages at 28 and 35 days p.t. were comparable in both treatment groups (99.6 and 99.3% for imidacloprid, 99.5 and 98.5 for fipronil, respectively). The figures declined slightly more in the fipronil-treated group (Hopkins 1997a).

Concentrating on a comparative evaluation of the speed of flea kill, Everett et al. (2000a) examined 36 artificially infested dogs, treated with imidacloprid (Advantage®), selamectin (Revolution™, Stronghold®, Pfizer), or left as untreated control. Evaluation was conducted at six, twelve, 24, and 36 hours after treatment and reinfestation. This was repeated on a weekly basis. The dogs were treated at the minimum labeled dose for the two products (imidacloprid 10 mg/kg b.w., selamectin 6 mg/kg b.w.). Imidacloprid provided significantly greater flea kill than selamectin six, twelve, and 24 hours after treatment and at six and twelve hours after every reinfestation. It further achieved significantly greater 36-hour flea kill at days 34 and 41 after treatment. At the 6-hour p.t./reinfestation flea count, imidacloprid was superior relative to selamectin on study days 0 through 27. Selamectin provided significant 6-hour flea control (relative to controls) only on day 13. At twelve hours after treatment/reinfestation, imidacloprid provided significant flea control (compared with controls) throughout the study (days 0 through 41). Imidacloprid was significantly superior to selamectin in all 12-hour samples with the exception of day 20. Selamectin provided 12-hour flea control significantly better than the untreated group on days 6, 13, and 20. At 24 hours after treatment, imidacloprid provided significantly better control than selamectin (97.6% versus 13.8%). By test day 6 and continuing through day 41 both treatments provided statistically equivalent 24-hour flea control. Finally, at 36 hours p.t./reinfestation, both treatment groups had significantly fewer fleas than controls on days 0 through 34, and no statistical differences between the efficacy of the two compounds through day 27. For days 34 and 41, however, imidacloprid provided significantly better 36-hour flea control than selamectin (Everett et al. 2000a, 2000b). In summary, both products provided significant flea control relative to untreated controls for at least one month after application. Imidacloprid provided significantly greater flea kill at six, twelve, and 24 hours p.t., a superiority which was maintained at the 6- and 12-hour flea counts throughout the study on all reinfestation days, except the 12-hour flea count on day 20. Both products provided measurable flea control beyond 30 days, with imidacloprid being superior in control relative to selamectin at the 36-hour flea counts on days 34 and 41 (Everett et al. 2000a). A highly significant flea control in as little as six hours after application was provided by imidacloprid, imidacloprid was generally significantly more rapid in flea kill compared with the results provided by selamectin (Everett et al. 2000a) (Table 28, Table 29, Table 30, Table 31).

Table 28. Comparison of the speed of flea kill between imidacloprid and selamectin: percentage of flea control six hours after treatment/reinfestation

| Day p.t. | | Advantage® (imidacloprid) | Revolution™ (selamectin) | Control |
|---|---|---|---|---|
| 0 | x Fleas | 9.2 | 79.6 | 68.9 |
| | % Control | 86.6[a,b] | −15.4 | - |
| 6 | x Fleas | 1.4 | 38.1 | 94.3 |
| | % Control | 98.5[a,b] | 56.6 | - |
| 13 | x Fleas | 1.0 | 40.5 | 87.8 |
| | % Control | 98.9[a,b] | 53.9[a] | - |
| 20 | x Fleas | 7.3 | 48.0 | 88.4 |
| | % Control | 91.7[a,b] | 45.6 | - |
| 27 | x Fleas | 3.7 | 48.2 | 77.8 |
| | % Control | 95.3[a,b] | 38.0 | - |
| 34 | x Fleas | 43.8 | 61.2 | 73.0 |
| | % Control | 40.0 | 16.1 | - |
| 41 | x Fleas | 31.8 | 61.5 | 84.9 |
| | % Control | 62.5 | 27.6 | - |

[a] significantly different from control (P < .05)
[b] significantly different from Revolution™ (P < .05)

Table 29. Comparison of the speed of flea kill between imidacloprid and selamectin: percentage of flea control twelve hours after treatment/reinfestation

| Day p.t. | | Advantage® (imidacloprid) | Revolution™ (selamectin) | Control |
|---|---|---|---|---|
| 0 | x Fleas | 2.7 | 72.9 | 82.0 |
| | % Control | 96.7[a,b] | 11.0 | - |
| 6 | x Fleas | 1.7 | 12.3 | 84.4 |
| | % Control | 98.0[a,b] | 85.4[a] | - |
| 13 | x Fleas | 1.6 | 14.7 | 98.6 |
| | % Control | 98.4[a,b] | 85.1[a] | - |
| 20 | x Fleas | 2.1 | 6.3 | 80.4 |
| | % Control | 97.4[a,b] | 92.1[a] | - |
| 27 | x Fleas | 2.4 | 24.9 | 83.2 |
| | % Control | 97.1[a,b] | 70.1 | - |
| 34 | x Fleas | 3.5 | 28.6 | 71.6 |
| | % Control | 95.1[a,b] | 60.0 | - |
| 41 | x Fleas | 4.4 | 34.7 | 89.7 |
| | % Control | 95.1[a,b] | 61.3 | - |

[a] significantly different from control (P < .05)
[b] significantly different from Revolution™ (P < .05)

Table 30. Comparison of the speed of flea kill between imidacloprid and selamectin: percentage of flea control 24 hours after treatment/reinfestation

| Day p.t. | | Advantage® (imidacloprid) | Revolution™ (selamectin) | Control |
|---|---|---|---|---|
| 0 | x Fleas | 1.8 | 66.2 | 76.8 |
| | % Control | 97.6[a,b] | 13.8 | - |
| 6 | x Fleas | 1.0 | 2.6 | 98.4 |
| | % Control | 99.0[a] | 97.3[a] | - |
| 13 | x Fleas | 1.0 | 2.1 | 101.0 |
| | % Control | 99.0[a] | 97.9[a] | - |
| 20 | x Fleas | 1.0 | 1.9 | 85.2 |
| | % Control | 98.8[a] | 97.8[a] | - |
| 27 | x Fleas | 1.3 | 4.2 | 89.2 |
| | % Control | 98.6[a] | 95.3[a] | - |
| 34 | x Fleas | 3.4 | 14.4 | 86.1 |
| | % Control | 96.0[a] | 83.3[a] | - |
| 41 | x Fleas | 5.5 | 20.7 | 89.6 |
| | % Control | 93.8[a] | 76.9[a] | - |

[a] significantly different from control (P < .05)
[b] significantly different from Revolution™ (P < .05)

Table 31. Comparison of the speed of flea kill between imidacloprid and selamectin: percentage of flea control 36 hours after treatment/reinfestation

| Day p.t. | | Advantage® (imidacloprid) | Revolution™ (selamectin) | Control |
|---|---|---|---|---|
| 0 | x Fleas | 1.0 | 1.0 | 89.7 |
| | % Control | 98.9[a] | 98.9[a] | - |
| 6 | x Fleas | 1.0 | 1.0 | 92.2 |
| | % Control | 98.9[a] | 98.9[a] | - |
| 13 | x Fleas | 1.0 | 1.0 | 65.5 |
| | % Control | 98.5[a] | 98.5[a] | - |
| 20 | x Fleas | 1.0 | 1.4 | 60.1 |
| | % Control | 98.3[a] | 97.6[a] | - |
| 27 | x Fleas | 1.3 | 4.8 | 79.8 |
| | % Control | 98.7[a] | 94.0[a] | - |
| 34 | x Fleas | 1.6 | 10.6 | 82.5 |
| | % Control | 98.1[a,b] | 87.2[a] | - |
| 41 | x Fleas | 2.6 | 46.4 | 51.7 |
| | % Control | 94.9[a,b] | 10.3 | - |

[a] significantly different from control (P < .05)
[b] significantly different from Revolution™ (P < .05)

Further comparison between imidacloprid and the two insecticides fipronil and selamectin in dogs treated at the recommended dosage showed different feeding behavior and time of lethal onset (Mehlhorn 2000; Mehlhorn et al. 2001b). Thirty adult fleas in two clean covers of plastic Petri dishes were fastened with tape to clipped, hairless zones at the side of previously treated dogs and further on removed and examined either one hour or 24 hours later. On the imidacloprid-treated dog fleas began feeding when placed on the dog, but stopped within three to five minutes. Furthermore they started to show tetanic and trembling movements and died within the first hour of exposure. Fleas on the selamectin- and fipronil-treated dogs started feeding and continued to do so during the first hour after placement. They remained fully mobile and showed jumping movements. At 24 hours after exposure, all fleas in the second Petri dish of the imidacloprid-treated dog were dead, while in the dishes of the two other treated dogs, a few of the fleas remained mobile. Dead fleas in both groups were engorged with blood. So, fipronil and selamectin both allowed longer feeding periods than imidacloprid (Mehlhorn 2000; Mehlhorn et al. 2001b) Detailed histological changes of the examined fleas of these studies can be seen in Fig. 37, Fig. 38 and Fig. 39.

Comparing imidacloprid with a larger number of compounds in its efficacy against cat flea infestations in dogs, Ritzhaupt et al. (1999) infested 28 dogs artificially with *C. felis* and treated them at monthly intervals with imidacloprid, fipronil, lufenuron/milbemycin Sentinel®, Novartis), or selamectin. For the first phase of the study, dogs were housed in a flea-infested environment for three months. For the second phase (starting on day 91), they were challenged with 20 adult fleas on a weekly base until the end of the study. With the exception of the lufenuron/milbemycin-treated group flea numbers in all other

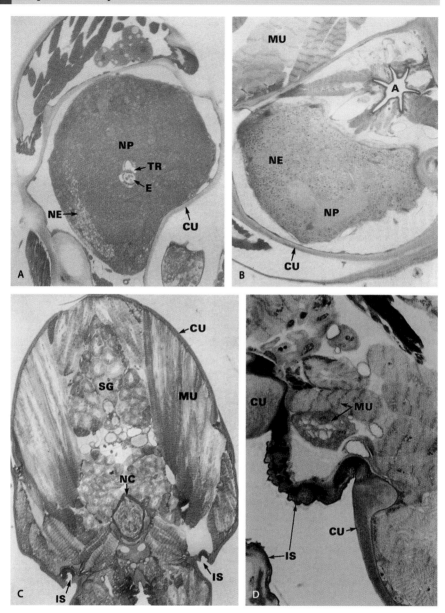

Fig. 37. A-D. Light micrographs of semithin sections through the anterior region of the head/body of adult fleas. **A** Oblique sections through the nerve system of an in vitro-treated stage, showing lightening of the nuclear region of nerve cells. Beside the central esophagus, tracheoles (TR) are present, but there is no alteration in the tissue. ×150. **B** Oblique sections through the subesophageal ganglion of an untreated flea; no degeneration is visible. ×150. **C** Sections through the thorax of a treated flea. Note the damage in the ventral nerve chord (NC) and along the nonsclerotized intersegmental membranes (IS). ×100. **D** Higher magnification of two intersegmental membranes (IS) within a treated adult flea; these membranes interconnect the thorax and the legs. Note that the muscles close to the membranes show degeneration. ×400

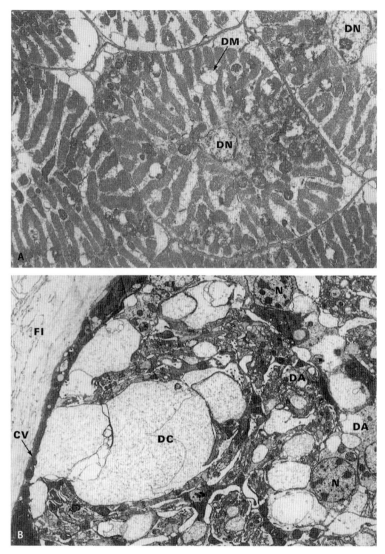

Fig. 38. A, B. TEMs of sections through an adult flea that had been exposed for one hour to the shaved skin of an imidacloprid-treated dog. Note the clear degeneration at the level of the cross-sectioned muscles (**A**) and along the periphery of the subesophageal ganglion (**B**). ×25,000

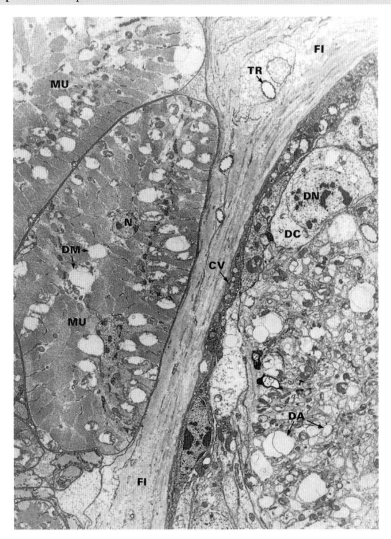

**Fig. 39.** TEM of a section through an adult cat flea that had been exposed in vitro for one hour to imidacloprid-impregnated filter paper. Note the extensive damage at the level of the muscle fibers and of the subesophageal ganglion. Most of the mitochondria and many axons were vacuolized (arrows). ×25,000

**Abbreviations for Fig. 37-39**
A Anterior stomach · CV Cellular cover of the ganglion · CU Cuticle · DA Degenerating axon · DC Degenerating nerve cell · DM Degenerating mitochondrion · DN Degenerating nucleus · E Esophagus with erythrocytes · FI Fibrillar layer of connective tissue · IS Intersegmental membrane · MU Muscle fiber · N Nucleus · NC Ventral nerve chord · NE Nerve cell · NP Neuropil (layer of axons) · SG Salivary glands · TR Tracheoles

test groups were reduced by 97 to 99% 14 days after treatment. The same compounds reduced mean flea burdens 99 to 100% from day 29 through to the end of the study. In the lufenuron/milbemycin-treated group geometric means of flea counts rose 87% (compared to day -6). Lufenuron(/milbemycin) reduced mean flea counts by 45% on day 29, and by 81 to 93% through day 150 (Ritzhaupt et al. 1999). This study clearly emphasizes the importance of concomitant adulticidal therapy.

A further comparative study between imidacloprid (Advantage®), fipronil (Frontline®), permethrin (Exspot®, Essex), and diazinon (Droplix®, Virbac) was performed on 25 dogs, artificially infested and reinfested weekly over six weeks after treatment (Liebisch and Reimann 2000). The study compared two compounds, already known for a long time (permethrin and diazinon), and two compounds quite recently introduced on the veterinary market (imidacloprid and fipronil). Flea counts were performed at 24, 48, and 72 hours after treatment and reinfestation. One hundred percent kill at 24 hours was observed with all products, 100% efficacy was further observed in all treatment groups until the evaluation after the third reinfestation. After the reinfestation three weeks post treatment, only the imidacloprid- and diazinon-treated groups were still free of fleas. Following the infestation on day 28, live fleas were recovered in all groups after 24 hours. Thereafter differences between the single groups became more and more evident, with imidacloprid showing highest efficacy. The reduction rate 24 hours after the reinfestation on day 28 was 98.96% and on day 30 again 100% in the imidacloprid-treated group. Permethrin was no longer sufficiently effective after day 28, while diazinon and fipronil were still clearly effective after the infestation on day 28, although the reduction had dropped below 95% for both. After day 35, permethrin had virtually lost its effect. On day 37 p.t. the flea reduction with permethrin was 55%, with fipronil 83%, with diazinon 86%, and with imidacloprid 99.8%. In summary, imidacloprid clearly had the longest duration of action with 37 days, showing excellent efficacy of 99.8% 37 days p.t.. The results may differ in a practical setting due to individual (coat length, animal size) and exogenous (keeping conditions) factors (Liebisch and Reimann 2000).

There was also a possible repellent effect of imidacloprid observed in the study, comparable to findings associated with permethrin products, applied as bath or spray. The suggestion is based on the absence of both live or dead fleas 48 hours p.t. in contrast to still dead fleas found on the coat of diazinon- and fipronil-treated dogs at the same time (Liebisch and Reimann 2000). This repellent effect could not be shown in studies of Mehlhorn (2000) respectively Mehlhorn et al. (2001b), testing the behavior of larval and adult fleas exposed to filter paper previously impregnated with imidacloprid or fipronil or selamectin. In all three groups no behavior was recorded, indicating that adult or larval fleas were attempting to avoid contact with the surface of compound-impregnated filter paper (Mehlhorn 2000; Mehlhorn et al. 2001b).

The efficacy of various insecticidal compounds was tested against an insecticide resistant strain of fleas titled 'cottontail'. Imidacloprid in a concentration of 5 mg/kg b.w. was the only adulticidal compound in in vivo tests with infested cats possessing 100% efficacy lasting over three weeks (Bardt and Schein 1996). Fipronil (Frontline® Spray) showed 40% efficacy over the period of two weeks. Under experimental conditions the strain was resistant to HCH, carbamates, organophosphates, rotenone, pyrethrin, pyrethroids, ivermectin, and lufenuron, but susceptible and sensitive to the developmental inhibitors triflumuron, and pyriproxyfen, as well as to methoprene (Bardt and Schein 1996). Imidacloprid-treated cats were free of fleas 24 hours after treatment and were protected for up to three weeks (Bardt and Schein 1996). In contrast, in studies by Pollmeier et al. (1999) the cottontail flea strain demonstrated susceptibility to fipronil with an efficacy over 95% throughout the study.

Comparative studies of imidacloprid and fipronil for the control of fleas on dogs shampooed pre-treatment or water immersed after treatment as well as fipronil in comparison to an amitraz collar solely or in combination with imidacloprid in water immersed dogs are reported in the chapter 'Effects of Shampooing and Repeated Water Exposure'.

## Field Studies

Off-host stages of fleas can be killed by direct application of chemicals in the domestic environment (MacDonald 1995; Logas 1995). An alternative approach is to progressively deplete this (domestic) reservoir by ensuring that fertile flea eggs are not deposited (Jacobs et al. 1997b). This can be achieved by carefully timed on-host treatments that either kill adult fleas before they start to produce eggs (Fisher et al. 1995) or render them infertile (Shipstone and Mason 1995).

Besides adulticidal compounds, there is another category of insecticidal compounds that exert their effect by interfering with the insect development. These compounds do not kill adults and thus have a delay in influencing insect populations (Kunkle 1997). Among these therapeutics, insect growth regulators (IGRs) mimic endogenous insect growth hormones whereas insect development inhibitors (IDIs) interfere with the insect's development in other ways (Kunkle 1997). Insect development inhibitors have no direct effect on adult fleas, so pets may continue to be infested with fleas for a considerable period if these products are used alone (Paul and Jones 1997).

The IDI lufenuron, a benzoylphenyl urea, is a chitin synthesis inhibitor (tested in systemical administration by Hink et al. (1991), Hink et al. (1994) and Blagburn et al. (1994)). It should be accompanied by an insecticide with adulticidal activity for the elimination of preexisting adult fleas, as the compound itself is only effective in preventing flea reproduction and larval development (Dryden et al. 1999b). Lufenuron reduces the viability of the next generation of flea eggs and larvae after the adult flea feeds on the blood of a treated animal. It prevents the development of flea eggs by inhibition of the synthesis, polymerization, and deposition of chitin, the major supportive component of the flea egg case and the exoskeleton of larvae stages (Paul and Jones 1997). It has no adulticidal action, but disrupts the development of immature stages by inhibiting the chitin synthesis. It acts systemically, so eggs produced by fleas taking blood from a treated host fail to hatch (Hink et al. 1991). Pyriproxyfen is one of the newest and most potent IGRs, possessing juvenile hormone activity (Kunkle 1997).

Even at low concentrations it has strong ovicidal and larvicidal properties (Kunkle 1997) and may have some delayed adulticidal effects (Blagburn and Lindsay 1995). In an on-animal-application it has the potential advantage that chemical residues are transferred from the coat of the animal to the blanket or other materials on which the animal lies and where developing flea stages are likely to be found. Thus pre-existing eggs and larvae are exposed to IGRs, as well as those derived from untreated animals (Fox 1996). Comparative evaluation of Advantage® (imidacloprid) and Program® (lufenuron, Novartis) for flea control on dogs was performed in a study by Paul et al. (1997) and Paul and Jones (1997), using 23 dogs in a controlled simulated home environment. The dogs were challenged with an artificial infestation and an untreated 'seeder' dog in each group. Both products were administered according to the manufacturers' recommendations at monthly intervals (Paul and Jones 1997; Paul et al. 1997). Thus the effect of the compounds on dogs with existing flea populations as well as the effect of exposure of treated dogs to an untreated, flea-infested, so-called 'seeder' dog was examined (Paul and Jones 1997). The efficacy in the imidacloprid-treated group ranged from 95.8-100%. In the lufenuron-treated group it ranged from 0-87.1% (Paul et al. 1997) (Table 32). Apart from the high efficacy of imidacloprid on the treated dogs, no live fleas were recovered from the seeder dog of this treatment group. Apparently, a sufficient quantity of imidacloprid was transferred from the treated dogs to the untreated ones, resulting in effective flea control (Paul and Jones 1997). The seeder dogs were rotated on a weekly basis to keep up the introduction of a fresh source of potential flea infestation in the imidacloprid-treated group. Small numbers of fleas were recovered on imidacloprid-treated dogs at days 21 and 28, but returned to zero after the treatment on day 28, lasting until day 56. After the third treatment on day 56, no further fleas were found on these dogs for the remainder of the study (Paul and Jones 1997). The lufenuron-treated dogs had flea counts similar to control animals through day 21, declining thereafter steadily until the end of the study (Paul and Jones 1997). In summary, the adulticidal activity of imidacloprid is demonstrated against existing flea populations. Monthly application reduces the flea counts immediately to zero, with only small numbers recovered at the end of two treatment periods. It further reduced or eliminated the fleas on untreated dogs with physical contact to the treated ones. Lufenuron did not show a decrease in flea counts for several weeks due to its lack of adulticidal activity, but showed a decline thereafter. Thus the necessity of a possible concurrent use of conventional insecticides, in the case of existing flea populations is illustrated (Paul

Table 32. Mean weekly flea counts on experimentally infested dogs treated with imidacloprid (group 1), lufenuron (group 2) or an untreated control group (group 3)

| | -7 | 1 | 7 | 14 | 21 | 28 | 35 | 42 | 49 | 56 | 63 | 70 | 77 |
|---|---|---|---|---|---|---|---|---|---|---|---|---|---|
| | | | | | | Study day | | | | | | | |
| Group 1 | 46.6 | 0 | 0 | 0 | 2.1 | 1.6 | 0 | 0 | 0 | 0.3 | 0 | 0 | 0 |
| Group 2 | 43.3 | 87.0 | 97.7 | 78.7 | 43.4 | 40.6 | 16.3 | 15.6 | 12.0 | 11.0 | 3.3 | 3.3 | 2.3 |
| Group 3 | 41.6 | 80.4 | 81.0 | 62.4 | 50.6 | 42.9 | 40.1 | 31.7 | 47.1 | 38.1 | 25.6 | 16.9 | 12.6 |

and Jones 1997). The use of an adulticide (e.g. pyrethrin spray) together with lufenuron versus imidacloprid alone is described below.

In cats, use of imidacloprid has been compared with lufenuron and pyriproxyfen under controlled simulated home environment conditions (Fox 1996; Jacobs et al. 1997b (comparison with lufenuron)). Cats kept and treated in groups of four were housed in a flea-infested area. After 56 days they were additionally infested with five extraneous fleas at weekly intervals (in cats with a flea burden below 30) to mimic roaming activities. Cats were treated with imidacloprid or lufenuron, both at 28 day intervals (Jacobs et al. 1997b), or at 56 day intervals in the case of pyriproxyfen, while a proportion of the animals was left as untreated controls (Fox 1996; Jacobs et al. 1997b). Welfare combings were required 18-times in the control pens after day 42 and in the lufenuron-treated cats three times, between days 100 and 105. At day 42 imidacloprid achieved 100% protection, pyriproxyfen 99.4%, and lufenuron 88.7%, respectively. No fleas were found at any time on the imidacloprid-treated animals, while small numbers (0-4) were subsequently found on pyriproxyfen-protected cats, and greater numbers, requiring the above mentioned welfare combings, on the lufenuron-protected ones (Fox 1996; Jacobs et al. 1997b). This points out the high efficacy of imidacloprid in flea control under simulated home conditions, and the better potential of pyriproxyfen compared to lufenuron in reaching pre-existing eggs and larvae. In the lufenuron group, flea numbers initially mirrored those of the controls, but by day 42, the ovicidal benefits became apparent, with 88.7% fewer fleas on treated cats. Later substantial flea burdens accumulated and three welfare combings were necessary after day 100 (Table 33). Correct use of lufenuron in the field (Shipstone and Mason 1995; Franc and Cadiergues 1995) or in experimental models (Fisher et al. 1995; Blagburn et al. 1995; Smith et al. 1996) would be expected to maintain lower flea populations (Jacobs et al. 1997b). Flea burdens below the limit of detection of the counting technique for the duration of the study (112 days) in the imidacloprid group could be due to any or all of the following three mechanisms (Jacobs et al. 1997b):

- too few flea eggs were possibly produced to maintain the life cycle
- residual insecticidal properties on the cat could have killed emergent fleas moving from the carpet onto treated cats
- trace amounts of a.i. from the cats' coat transferred to the carpet to exert a larvicidal effect on the developing flea population (Hopkins et al. 1996b)

All in all, a greater degree of protection was provided by the adulticide imidacloprid than by the regular use of an IGR alone (Jacobs et al. 1997b). Similar conclusions were also drawn in an earlier study by Fisher et al. (1995) comparing a less potent adulticide, fenthion, with lufenuron.

The study of Jacobs et al. (1997b) indicated that control of a flea population within a household may be possible with carefully timed on-host adulticidal treatments. But generally, regardless which compound is used, the treatment interval must lie within the period of high residual efficacy of the formulated product (Jacobs et al. 1997b). The results of Jacobs et al. (1997a), showing an initial efficacy of 100% in cats treated with a dose of 10 mg imidacloprid/kg b.w., which decreased to 96% four weeks after

Table 33. Individual and mean flea counts of three groups of cats: untreated, and treated at 28-day intervals with imidacloprid or lufenuron

| Treatment group | Cat no. | Flea counts on day | | | | | | | |
|---|---|---|---|---|---|---|---|---|---|
| | | 14 | 28 | 42 | 56 | 70 | 84 | 98 | 112 |
| Control | 1 | 8 | 6 | 94 | 73 | 46 | 46 | - | - |
| | 2 | 8 | 6 | 43 | 60 | 67 | 39 | 55 | 87 |
| | 3 | 2 | 10 | 50 | 60 | 38 | 37 | 58 | 104 |
| | 4 | 14 | 8 | 117 | 142 | 50 | 70 | 74 | 95 |
| | Mean | 6.0[a] | 7.4[a] | 69.8[a] | 78.2 | 49.2 | 46.5 | 61.8 | 95.1 |
| Imidacloprid | 5 | 0 | 0 | 0 | 0 | 0 | 0 | - | - |
| | 6 | 0 | 0 | 0 | 0 | 0 | 0 | 0 | 0 |
| | 7 | 0 | 0 | 0 | 0 | 0 | 0 | 0 | 0 |
| | 8 | 0 | 0 | 0 | 0 | 0 | 0 | 0 | 0 |
| | Mean | 0 | 0 | 0 | 0 | 0 | 0 | 0 | 0 |
| Lufenuron | 13 | 11 | 18 | 19 | 51 | 87 | 76 | - | - |
| | 14 | 13 | 6 | 6 | 24 | 13 | 14 | 54 | 36 |
| | 15 | 1 | 1 | 8 | 17 | 12 | 13 | 58 | 22 |
| | 16 | 5 | 3 | 4 | 6 | 5 | 13 | 59 | 43 |
| | Mean | 5.7[a] | 4.7[a] | 7.9[b] | 19.1 | 16.6 | 20.8 | 57.0 | 32.5 |

[a,b] = means with different superscripts are statistically different ($P < 0.05$)
- = no count made (cats removed from study on day 88 following an injury)
no statistical analysis performed after day 42 as flea burden of controls reduced on welfare grounds

treatment, suggest that monthly applications of imidacloprid during the flea season could reduce flea egg production to an epidemiologically inconsequential level (Jacobs et al. 1997b).

Imidacloprid was furthermore tested in comparison with a combination of lufenuron and pyrethrin spray in 37 dogs and 19 cats living in flea-infested households in the Tampa area, Florida (Dryden et al. 1999b). The precondition for participation in the study were at least three adult fleas collected in two lighted flea traps in the homes during a particular time period and a minimum of three adult fleas observed on at least one of the pets in the household (Dryden et al. 1999b). In one group of households the pets were treated with imidacloprid according to the manufacturer's directions at monthly intervals over the period of 90 days, in the other group the pets were given oral lufenuron also according to the label recommendations once a month for 90 days and were furthermore treated topically once every one to two weeks with a water-based pyrethrin spray if fleas could be detected on the animals (Dryden et al. 1999b). The initial application of imidacloprid resulted in a reduction of 96% in flea numbers on day 7 and between 93 to 96.5% reduction on days 14, 21, and 28. In the other group, the combination treatment reduced flea numbers on pets by 48.9% on day 7 and 91.1% on day 28. While the first application of lufenuron did not kill preexisting adult fleas, the decrease in numbers during the first few weeks was only attributable to the pyrethrin spray. Thus this insecticide provided only 48.9, 30.5, and 66.6% reduction in flea numbers on day 7, 14, and 21 respectively (Dryden et al. 1999b). The inadequate flea control provided by the pyrethrin spray was either attributable to a lack of residual activity of the formulation or to pyrethrin-resistant fleas (Dryden et al. 1999b). Apart from the numbers on day 28, all flea counts from day 7 to day 40 to 45 were significantly different between the two groups (Dryden et al. 1999b) (Table 34). Furthermore the environmental flea counts via light traps did not show any marked decrease from pre-treatment values during the first two weeks p.t.. On days 21 and 28 however flea burdens were reduced, compared to the beginning of the treatment (day 0), by 74.8 and 86.8% in the imidacloprid group and by 25.3 and 44.0% in the lufenuron-pyrethrin group. At any time the mean flea numbers trapped were not significantly different between the two groups (Dryden et al. 1999b) (Table 35). Furthermore it was demonstrated that flea control can be achieved solely by effective topical or systemic administration of insecticides without concomitant treatment of the surroundings.

By the end of the study, following three applications of imidacloprid, flea burdens on pets and in homes were reduced by 98.8 and 99.9%, respectively. A reduction of 99.2 and 99.7% on pets and in homes was also achieved at the end of the study with the treatment combination lufenuron and pyrethrin spray (Dryden et al. 1999b).

In another study in Tampa, Florida, efficacy comparisons between topical spot applications of 9.1% w/v imidacloprid and 9.7% fipronil in 20 flea-infested homes housing 23 dogs and nine cats revealed 95.5 and 97.5% efficacy in reducing flea populations on pets at seven and 28 days after a single application of imidacloprid and furthermore 97.6 and 97.1% reduction at the same time points after a single application of fipronil (Dryden et al. 1999a, 2000). After three months of treatment with imidacloprid or fipronil, flea burdens on pets were reduced by 99.5 and 96.4%, respectively. Flea burdens in the environment, assessed using intermittent light traps, were reduced by 100.0% and 98.8% in homes where pets were treated with imidacloprid or fipronil,

Table 34. Mean (± SD) number of fleas observed on pets* in flea-infested households after pets were treated with imidacloprid applied topically once a month (19 households) or lufenuron administered orally once a month and pyrethrin spray applied topically once every one to two weeks (15 households)

| Days p.t. | Imidacloprid (dogs, n = 19; cats, n = 14) | | Lufenuron and pyrethrin spray (dogs, n = 18; cats, n = 5) | |
|---|---|---|---|---|
| | No. of fleas | Mean % reduction | No. of fleas | Mean % reduction |
| 0 | 16.6 ± 13.7 | 0.0 | 26.4 ± 30.6 | 0.0 |
| 7 | 0.5 ± 1.4 | 96.0[a] | 8.3 ± 11.3 | 48.9 |
| 14 | 1.4 ± 2.5 | 93.0[a] | 11.3 ± 17.7 | 30.5 |
| 21 | 0.5 ± 1.0 | 96.5[a] | 8.5 ± 17.8 | 66.6 |
| 28 | 1.0 ± 1.6 | 93.6 | 2.3 ± 3.5 | 91.1 |
| 40-45 | 0.03 ± 0.2 | 99.4[a] | 3.7 ± 8.0 | 83.3 |
| 54-60 | 0.0 ± 0.0 | 100.0 | 2.5 ± 8.2 | 94.1 |
| 69-75 | 0.03 ± 0.2 | 99.8 | 0.9 ± 2.3 | 96.1 |

* detected by visual inspection for one minute after parting hair at the dorsal midline, tailhead, and left and right lateral thoracic and inguinal regions of each pet
[a] significantly (P < 0.05) different from mean percentage reduction determined at the same period for pets treated with lufenuron and pyrethrin spray

Table 35. Mean (± SD) number of fleas collected in flea-infested homes* after pets were treated with imidacloprid applied topically once a month (19 households) or lufenuron administered orally once a month and pyrethrin spray applied topically once every one to two weeks (15 households)

| Days p.t. | Imidacloprid | | Lufenuron and pyrethrin spray | |
|---|---|---|---|---|
| | No. of fleas | Mean % reduction | No. of fleas | Mean % reduction |
| 0 | 43.5 ± 75.3 | 0.0 | 59.9 ± 80.3 | 0.0 |
| 7 | 30.3 ± 60.5 | 4.7 | 24.6 ± 5.6 | 10.7 |
| 14 | 33.1 ± 83.7 | 19.8 | 35.4 ± 67.3 | 19.9 |
| 21 | 10.5 ± 27.4 | 74.8 | 42.7 ± 84.8 | 25.3 |
| 28 | 2.2 ± 4.9 | 86.8 | 28.8 ± 65.2 | 44.0 |
| 40-45 | 0.2 ± 0.3 | 99.0 | 4.5 ± 5.7 | 85.8 |
| 54-60 | 0.1 ± 0.2 | 99.9 | 0.8 ± 1.2 | 97.4 |
| 69-75 | 0.1 ± 0.2 | 99.9 | 0.5 ± 1.7 | 98.9 |

* collected in intermittent light traps placed in two rooms of each home for 16 to 24 hours

respectively. Thus both topical spot applications markedly reduced flea infestations in naturally infested homes without the use of environmental treatments (Dryden et al. 1999a, 2000).

# Larvicidal Effect of Imidacloprid and Efficacy Enhancement (with PBO)

## Larvicidal Effect

Apart from the nearly immediate, strong, and effective adulticidal activity of imidacloprid on on-host populations of fleas, observations and investigations concerning a larvicidal effect of imidacloprid have been performed. Reinfestation of an animal from its surroundings on the basis of earlier deposited eggs, developed larvae, and preemergent adults can either be overcome by the residual activity of an adulticide or by repeated treatments until the reinfestation stops, possibly taking some weeks (Hopkins et al. 1997b). This is provided that the adulticide is effective before female fleas start producing fertile eggs, which means during the first 36 hours after a flea infests an animal (Osbrink and Rust 1984). A further possibility to stop reinfestation is the reduction or elimination of immature flea stages in the surroundings, either by vacuuming and other physical means or by the use of effective larvicidal compounds (Hopkins et al. 1997b).

A first study to determine the presence of any larvicidal effect of imidacloprid was conducted by Hopkins et al. (1997b). Imidacloprid 10% w/v topical formulation was tested in the immediate surroundings of treated dogs. Debris falling from treated dogs were collected on different days and subsequently added to Petri dishes in which C. felis eggs had been incubated to develop into first instar larvae, testing the effect of the debris on the viability of larvae. The larvicidal effect was calculated by comparing the number of live larvae in contact with debris from treated or untreated dogs, while the calculated percentage mortality in the treated samples was corrected by the natural mortality in the untreated group (using a formula of Tattersfield and Morris 1924) (Hopkins et al. 1997b). In a second study the practical significance of any larvicidal effect was assessed by measuring the development of fleas to the pupal and adult stage in a room exposed to imidacloprid-treated dogs, compared with an unexposed room. The rooms were seeded with flea eggs from untreated dogs and then only one of the rooms meanwhile had contact with imidacloprid-treated dogs. Sawdust samples of both rooms, taken at different days and artificially seeded with flea eggs, were tested as medium for the development of immature fleas. Later on the samples were also examined for immature development of naturally seeded eggs. Finally both rooms were checked for adult fleas by the so-called 'white sock technique', with a person dressed in white socks and overall, moving around in each room in order to detect fleas jumping onto the socks and the legs of the overall (Hopkins et al. 1997b). The effect of the treated dogs on the immature stages in their surroundings was assessed by comparing the numbers of larvae and pupae found in identical random samples examined, as well as by comparing the number of adult fleas found in a specified time (Hopkins et al. 1997b).

In the first study, debris from the treated dogs inhibited larval development by an average of more than 99% over four weeks after treatment. Even about two months later the effect of the debris collected on days 2 and 7 was still evident (Hopkins et al. 1997b). Detailed percentages of inhibition in samples of different days can be seen in Table 36. No pupae developed in any sample from treated dogs, whereas pupation occurred in control samples. At all times most larvae were dead or affected within four hours of application (Hopkins et al. 1997b).

In the second study significant differences between both rooms were recorded at all time intervals (Hopkins et al. 1997b). Floor dust samples, artificially seeded with eggs, showed an inhibition of larval and pupal development due to exposure to imidacloprid-treated dogs between 91 and 99%. Floor dust samples of the rooms, which had been 'naturally' seeded by infested dogs and later on examined for development, showed 87% inhibition of larval and 100% inhibition of pupal development in the samples of day 29 p.t.. Using the white sock technique 94 adult fleas were found within 14 seconds in the untreated room whereas only three fleas could be recorded after twelve minutes in the room which had been exposed to treated dogs. An inhibition of 97% due to imidacloprid was recorded. The overall average effect of exposure to imidacloprid-treated dogs, based on pupal and adult flea numbers, was 98.6% reduction in development to adult fleas in the surroundings. All in all, a strong larvicidal effect could be observed testing debris from imidacloprid-treated dogs, even though the quality and quantity of debris falling from treated dogs may vary, depending on breed, coat type, activity, effect of water in the coat, etc.. Hair had been excluded from the debris. So mainly small particle debris of the type falling from imidacloprid-treated dogs, such as epidermal scales, was examined. This material is likely to follow the same distribution as flea feces, the main food source of larvae, and therefore provides a more thorough distribution of effective larvicidal activity (Hopkins et al. 1997b).

Table 36. Mortality of *C. felis* larvae in contact with 5 to 10 mg of debris falling in one hour from dogs at various times after treatment with imidacloprid

| | Days p.t. | | | | | | Retested samples Day 7 samples 51 days later | Day 2 samples 61 days later |
|---|---|---|---|---|---|---|---|---|
| | 1 | 2 | 7 | 14 | 21 | 28 | | |
| Percent mortality | 100 | 98.8 | 100 | 100 | 95.3[b] | 100 | 100 | 97.4[c] |
| | | No pupae[a] | | | No pupae | | | No pupae |

Overall inhibition of larval development > 99% compared to control samples.
[a] pupation occurred in control samples but not in samples with debris from treated dogs
[b] larvae were showing toxicosis but not yet dead
[c] larvae showing toxicosis; number is not corrected for escape of some larvae in one control sample

In the second study by the time development had reached pupal or adult stages, 98.6% fewer pupae or adults were found in the surroundings accessed by the treated dogs (Hopkins et al. 1997b). A release of small particle debris was continuous for at least four weeks. Important in this context is, that the distribution of this larvicidal debris must correlate to the animals' habitual movement patterns in the same way as the concentrations of flea eggs. Therefore this could be a very efficient method of control of flea reinfestations as the inhibition will be closely correlated to the spatial distribution of eggs and larvae and will not be wasted on areas where the animal does not stay (Hopkins et al. 1997b).

With the larvicidal potential, imidacloprid may provide a form of environmental control which concentrates in the areas most frequented by the dog or cat.

The effect of clipped hair (seven days after treatment) of imidacloprid-, fipronil-, or selamectin-treated dogs on adult and larval fleas was tested by Mehlhorn (2000) respectively Mehlhorn et al. (2001b). In the case of imidacloprid all adult fleas died within 50 to 90 minutes. Using fipronil-treated hair most adult fleas died within 24 to 29 hours, but some fleas remained mobile and survived. The majority of adult fleas added to hair from the selamectin-treated dog died within 96 hours. But several fleas survived and remained mobile beyond this period, as did fleas of the control group in general (Mehlhorn 2000; Mehlhorn et al. 2001b). Flea larvae exposed to hair from dogs treated seven days previously survived for six hours when placed on clipped hair from the imidacloprid-treated dog. In the case of fipronil and selamectin, flea larvae survived for 29 hours (Mehlhorn 2000; Mehlhorn et al. 2001b).

The need for environmental treatments with larvicidal chemicals is probably eliminated, even though full realization of the larvicidal effect is dependent on the owner not restricting the animal's normal movement patterns in its surroundings, especially its normal sleeping and resting areas. One hundred percent control of immature stages cannot be expected. But in combination with the strong adulticidal effect of imidacloprid any treatment of the surroundings for control of immature flea stages is probably not necessary (Hopkins et al. 1997b).

Comparable to the target of the above mentioned dog trial of Hopkins et al. (1997b), the larvicidal effect of imidacloprid has also been evaluated for the control of fleas in the environment of treated cats (Liebisch and Krebber 1997). The larvicidal activity of imidacloprid on flea larvae was tested by incubating flea eggs on parts of blankets onto which imidacloprid-treated cats had been placed for six hours daily. Furthermore the amounts of imidacloprid, released from imidacloprid-treated cats, were evaluated by extracting and measuring them out of blanket parts according to a laboratory protocol. The kinetic of imidacloprid contents in mg/m² blanket can be seen in Table 37. Blankets were renewed weekly during the study. The main part of imidacloprid found in the blankets had been released during the first six hours of contact. Additional time of contact did not lead to a linear increase of the amount of imidacloprid on the blankets. The effect of imidacloprid as a larvicide in frequently used parts of the animal's environment is expressed in percentages of reduction of flea offspring development on parts of blankets which have been in contact with treated cats for varying amounts of time at different times after treatment (Liebisch and Krebber 1997) (Fig. 40). After 18 hours of contact (on day three) no development of fleas into adulthood could be observed in the treated group, in contrast to an average of 24.4 fleas on blanket samples of the untreated group (Liebisch and Krebber 1997). After a renewing of the blankets the numbers of fleas developing on the blanket samples increased, but were still below the numbers in the untreated group samples and again showed a decrease after the time interval to contact. The same tendency was observed in the two following weeks. After three weeks and 42 hours of contact with treated cats the blanket samples exhibited an efficacy of 74.7% reduction in number of developing fleas. After four weeks and 42 hours there was still an efficacy of 57.0% (Liebisch and Krebber 1997). The cat trial verified a significant larvicidal activity of imidacloprid in the surroundings of treated animals. The activity lasted at least three weeks after treatment. The grade of efficacy depended on the duration of contact between cats and the larval habitat and the time interval since treatment. The main part of the a.i. found in the blankets had been released during the first week, especially during the first six hours of contact (Liebisch and Krebber 1997).

The apparently complete suppression of the flea population obtained in imidacloprid-treated cats housed under simulated home environment conditions with

Table 37. Imidacloprid contents in mg/m² blanket (weight of blanket = 200 g/m²)

| Cat no. | '1 | "3 | '''7 | '8 | "10 | '''14 | '15 | "17 | '''21 | '22 | "24 | '''28 |
|---|---|---|---|---|---|---|---|---|---|---|---|---|
| | | | | **Days following treatment** | | | | | | | | |
| 1 | 6,34 | 6,26 | 3,64 | 0,1 | 0,1 | 0,76 | 0,1 | 0,14 | 0,36 | 0,1 | 0,1 | 0,3 |
| 2 | 2,52 | 5,14 | 8,98 | 0,1 | 0,1 | 0,56 | 0,1 | 0,1 | 0,14 | 0,1 | 0,1 | 0,22 |
| 3 | 2,96 | 3,6 | 5,76 | 0,44 | 0,42 | 1,14 | 0,1 | 0,1 | 0,18 | 0,1 | 0,1 | 0,24 |
| 4 | 8,56 | 5,38 | 9,86 | 0,1 | 0,2 | 0,32 | 0,1 | 0,1 | 0,22 | 0,1 | 0,1 | 0,28 |
| 5 | 5,25 | 2,12 | 6,64 | 0,1 | 0,42 | 0,54 | 0,1 | 0,12 | 0,32 | 0,1 | 0,14 | 0,1 |
| 6 | 11,06 | 15,62 | 11,82 | 0,42 | 0,86 | 0,58 | 0,1 | 0,2 | 0,34 | 0,1 | 0,16 | 0,1 |
| Mean | 5,66 | 6,36 | 7,78 | 0,18 | 0,34 | 0,66 | 0,1 | 0,12 | 0,26 | 0,1 | 0,1 | 0,2 |

' = 6 hours contact of imidacloprid-treated cats on blankets
" = 18 hours contact of imidacloprid-treated cats on blankets
''' = 42 hours contact of imidacloprid-treated cats on blankets

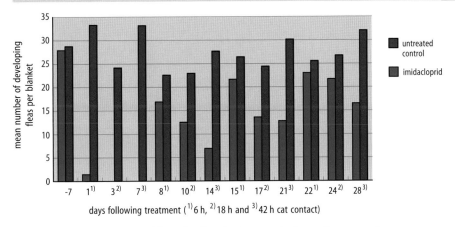

Fig. 40. Efficacy in reduction of flea offspring development in the environment of cats

an established flea population as shown by a control group (Jacobs et al. 1997b, reported above) exceeded the researchers' expectations. It had been assumed that treated cats would pick up some fleas in the early stages of the trial while the carpeted pens were still infested with off-host developmental stages (Jacobs 2000). A possible explanation for these results is provided by Hopkins et al. (1996b, 1997b) with their observation of larvicidal properties in the skin debris from treated dogs. This suggested that imidacloprid may have a short-term direct environmental effect as well as the longer-term impact obtained by stopping flea egg production (Jacobs 2000).

Ascertaining whether imidacloprid treatment of cats influences the development of flea larvae in their immediate environment as it has been observed with skin debris from imidacloprid-treated dogs in Petri dishes (Hopkins et al. 1996b, 1997b) another cat study was conducted by Jacobs et al. (2000): Imidacloprid-treated cats were allocated individually to a confined cage for six hours a day for five days a week from one week before treatment to four weeks after treatment. The blanket which covered the floor of every cage was changed weekly, and triplicated samples of these blankets were cut off and incubated with flea eggs and larval food following standardized procedures. Even though the fleece blankets did not provide ideal rearing conditions for flea development, highly significant differences were observed and calculated at each time between arithmetic means for percentage yield of adult fleas in treatment and control groups (Fig. 41). Flea viability on blankets used by treated cats was reduced by 100, 84, 60, and 74% in the first, second, third, and fourth week after treatment, respectively, compared with the number of adult fleas developing on the corresponding control blankets. Thus sufficient insecticide transferred from the treated cats to their blanket to exert a subsequently lethal effect on developing off-host life cycle stages (Jacobs et al. 2000). During the first week after treatment, when the lethal effect was maximal, the presence of pupae could not be demonstrated, indicating that mortality occurred primarily during the larval phase of development.

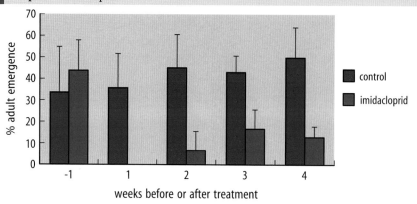

Fig. 41. Percentage of adult flea emergence (with standard deviations). Control group: eggs incubated on blankets used by untreated cats. Treated group: eggs incubated on blankets used by cats treated with imidacloprid.

So imidacloprid proved to exert significant larvicidal activity in the immediate environment of treated cats (Jacobs et al. 2000), which is likely to contribute to the rapid control of the flea population observed in the simulated environmental conditions of the trials of Jacobs et al. (1997b).

The conditions of the above summarized study of Jacobs et al. (2000) are artificial and cannot be directly extrapolated to a domestic situation, among other things because of the fact that cats may not be so closely confined under normal conditions. But nevertheless, the highest concentrations of flea eggs are found where pets spend most time (Rust and Dryden 1997). Cats spend several hours each day resting or sleeping, often at habitual sites, so the larvicidal effect will have a significant impact on flea development in the immediate vicinity (Jacobs et al. 2000). The larvicidal activity is thereby targeted to where it is most needed (Jacobs 2000). Furthermore, the larvicidal activity will also be expected in places where high concentrations of eggs are associated with energetic activity by the cat, such as jumping and landing (Robinson 1995), if imidacloprid attaches to epidermal scales, as observed in dogs (Hopkins et al. 1996b). In these areas of energetic activity skin debris would as well be deposited (Jacobs et al. 2000). In the study of Jacobs et al. (2000) blankets had been changed regularly on a weekly basis, to mimic owners laundering the pet basket each week. In reality, blankets tend to be changed irregularly and often less frequently, besides other habitual resting sites in the home or in outbuildings, which are cleaned less regularly. So the possibility of the initial high level of larvicidal activity to persist for a longer period is likely.

Unpublished data of ongoing studies to mimic situations closer to real domestic life indicate 98 to 100% reduction in the number of fleas in the preliminary results of in vitro tests (Jacobs 2000). In these trials blankets are used for a longer period of time in order to test whether accumulations of imidacloprid could provide greater larvicidal efficacy. The time spent on the blankets each day is again six hours on five days each week, but one treated group uses the blankets over a period of ten days, while the second even uses it over 20 days (Jacobs 2000).

All in all it is likely, that the larvicidal effect of imidacloprid is of practical significance in reducing the level of flea infestation in the domestic environment during the early stages of control programs using this compound (Jacobs 2000) (Table 38, Table 39 and Box 9). Without some form of environmental control animals may be exposed to reinfestation with fleas for several weeks after the initiation of a control program, reflecting the time taken for the completion of the off-host phase of the life cycle and the delayed emergence of adults as observed by Silverman and Rust (1985) (Jacobs et al. 2000).

Via epidermal scales of treated animals a larvicidal effect in the environment especially in the resting and sleeping areas exists. This effect differs depending on breed, hair coat length, type and moisture in the epidermis, and is mainly effective if vacuuming and cleaning procedures do not remove the scales.

Table 38. First larvicidal study: percentage of reduction of adult flea emergence[a]

| Week | Percentage of reduction |
|------|-------------------------|
| 1 | 100 |
| 2 | 84 |
| 3 | 60 |
| 4 | 74 |

[a] Flea eggs were incubated on blankets used by cats during the first, second, third, or fourth week after a single treatment with imidacloprid (compared with blankets used by untreated cats). All differences between control and treatment groups were significant (P <0.001).

Table 39. Second larvicidal study: number and percentage of reduction of adult fleas emerging[a]

| Period of use | Number of fleas | Percentage of reduction |
|---------------|-----------------|-------------------------|
| Control | 41 | — |
| 10 days | 0 | 100 |
| 20 days | 1 | 98 |

[a] Flea eggs incubated on blankets used by untreated controls or by imidacloprid-treated cats.

**BOX 9. Principle of environmental flea control by means of animal treatments**

**Breaking the flea life cycle:**

**If fleas on pets in a home are adequately controlled**
⬇
**No eggs will drop into environment**
⬇
**Reservoir of off-host life cycle stages will be progressively depleted until**
⬇
**No more reinfestation**

## Efficacy Enhancement

The insecticidal effect of imidacloprid on fleas investigated at different temperatures showed highest potency of the compound in adult cat fleas at 35°C and in larvae at 20°C (Richman and Koehler 1997; Richman et al. 1999). The use of the synergist piperonyl butoxide (PBO) increased the relative potency of imidacloprid (1:5 imidacloprid:PBO) ~16-fold at 26°C and ~5.5-fold at 30°C, but not at 35°C (Richman and Koehler 1997; Richman et al. 1999). In larvae doubled toxicity by 1:5 imidacloprid:PBO at 26°C was observed, but no synergism occurred at 20°C (Richman and Koehler 1997; Richman et al. 1999) (Table 40).

Table 40. Results of probit analyses for adult cat fleas exposed to imidacloprid and PBO at various temperatures

| Imidacloprid: PBO | Temp. °C | $n$ | Slope ± SD | $LC_{50}$[a] (95% CI) | $x^2$ | Relative potency[b] (95% CI) |
|---|---|---|---|---|---|---|
| 1:0 | 20 | 540 | 2.97 ± 0.28 | 0.22 (0.20-0.23) | 0.67 | — |
|  | 26 | 360 | 2.42 ± 0.24 | 0.32 (0.28-0.37) | 1.44 | — |
|  | 30 | 270 | 2.54 ± 0.35 | 0.20 (0.17-0.24) | 0.03 | — |
|  | 35 | 360 | 2.89 ± 0.29 | 0.05 (0.05-0.06) | 2.60 | — |
| 1:1 | 20 | 540 | 4.38 ± 0.31 | 0.20 (0.19-0.21) | 7.50 | 1.09 (0.97-1.23) |
|  | 26 | 510 | 1.59 ± 0.19 | 0.12 (0.10-0.14) | 1.44 | 2.60 (2.12-3.20) |
|  | 30 | 390 | 2.25 ± 0.30 | 0.12 (0.11-0.14) | 2.94 | 1.72 (1.41-2.10) |
|  | 35 | 360 | 1.72 ± 0.22 | 0.03 (0.03-0.04) | 0.74 | 1.40 (1.12-1.74) |
| 1:5 | 20 | 360 | 3.02 ± 0.27 | 0.04 (0.03-0.04) | 2.48 | 5.79 (5.05-6.64) |
|  | 26 | 360 | 2.26 ± 0.24 | 0.02 (0.02-0.02) | 0.31 | 16.17 (13.26-19.87) |
|  | 30 | 270 | 3.15 ± 0.41 | 0.03 (0.03-0.04) | 1.63 | 5.66 (4.47-6.98) |
|  | 35 | 270 | 2.89 ± 0.36 | 0.04 (0.04-0.05) | 0.56 | 1.07 (0.84-1.34) |

Chi-square values were not significant ($P < 0.05$)
[a] $\mu g/cm^2$
[b] $LC_{50}$ imidacloprid/$LC_{50}$ imidacloprid + PBO at same temperature
CI: confidence interval

# Environmental and Habitual Factors Influencing Imidacloprid Treatment

## Effects of Shampooing and Repeated Water Exposure

The effect of water exposure of treated animals on the residual efficacy of imidacloprid, with or without shampooing, is important for the establishment of a continuous flea control and has been the subject of a number of studies. Any product on skin surface or in skin fat will be removed by effective shampooing. Water immersion is far less significant.

Twenty-four flea-infested dogs of different breeds and a variety of hair coats were randomly assigned to three treatment groups. On day 4 after treatment 16 treated animals were either shampooed with a mild emollient shampoo (Allergoom® - Allerderm, Virbac) and rinsed on day 4 after treatment, or immersed in a tank of water for one minute to simulate swimming or exposure to heavy rainfall, a procedure which was repeated on days 11, 18, 25, and 32 (Arther et al. 1997a; Cunningham et al. 1997b). Similar treatment was administered to untreated control dogs. Flea counts were performed one day after treatment, respectively after every reinfestation (with 100 adult fleas), and percent flea control was determined by comparing the number of live fleas on treated and untreated dogs (Arther et al. 1997a; Cunningham et al. 1997b). Single shampoo/rinse procedure four days after treatment with Advantage® did not significantly decrease the residual flea control for up to four weeks. Efficacy following repeated infestation remained above 90% for three of the four flea count periods through day 28 (Arther et al. 1997a; Cunningham et al. 1997b) (Fig. 42). Immersion of treated dogs at weekly intervals, simulating swimming or rainfall, had as well negligible effect on the residual efficacy over 28 days (Arther et al. 1997a; Cunningham et al. 1997b)

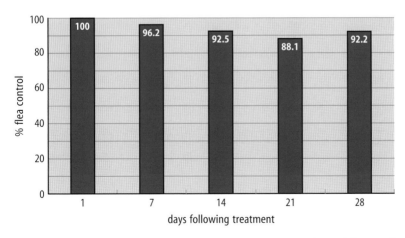

Fig. 42. Duration of flea efficacy of imidacloprid 10% spot-on-treated dogs following shampooing at day 4 p.t. and weekly reinfestation with 100 adult fleas

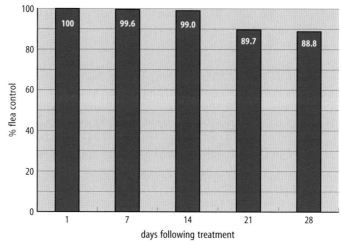

Fig. 43. Duration of flea efficacy of imidacloprid 10% spot-on-treated dogs after weekly water immersion and weekly reinfestation with 100 adult fleas

(Fig. 43). So, high level of flea control remains even if dogs are shampooed or if they swim or are exposed to rainfall between monthly treatments with imidacloprid 10% spot-on (Cunningham et al. 1997b). Detailed percentages can be found in Table 41.

Pre-treatment shampooing or post-treatment water immersion for one minute and their possible effects on flea control were also examined by Cunningham et al. (1999) in infested and weekly reinfested dogs either treated with fipronil (Frontline® Top Spot™) or imidacloprid as flea adulticides. Shampooing one hour before treatment did not affect flea efficacy in both treatment groups. Post-treatment water immersion for one minute on days 3, 6, 13, 20, and 27 did not affect the control efficacy in the fipronil group, showing significantly lower flea counts and fewer numbers of dogs infested than in the imidacloprid group. Treatment with imidacloprid revealed some infestation on days 17, 24, 31, and 45. Nevertheless the trials of Arther et al. (1997a) and Cunningham et al. (1997b) revealed a minimal impact on the residual efficacy of imidacloprid 10% spot-on over four weeks (see above).

Table 41. Mean number of fleas (% control) in imidacloprid-treated dogs after shampooing or multiple water exposure

| | | Days p.t. | | | | | |
|---|---|---|---|---|---|---|---|
| | Treatment/handling | 1 | 7 | 14 | 21 | 28 | 35 |
| Group 1 | imidacloprid-treated/ shampooed | 0.0 (100) | 3.4 (96.2) | 7.1 (92.5) | 11.4 (88.1) | 7.9 (92.2) | 30.8 (65.5) |
| Group 2 | imidacloprid-treated/ water immersed | 0.0 (100) | 0.3 (99.6) | 0.9 (99.0) | 9.1 (89.7) | 9.5 (88.8) | 19.8 (78.7) |
| Group 3 | control/shampooed control/water immersed | 95.0 67.1 | 90.5 76.5 | 95.0 87.3 | 95.8 88.0 | 100.5 84.5 | 89.3 92.8 |

The efficacy of imidacloprid 10% spot-on in combination with an amitraz (Preventic®, Virbac) collar against fleas and ticks, compared to fipronil (Frontline® Top Spot™) or an amitraz collar alone was tested after water immersion (for one minute), in adult flea- and tick (*Rhipicephalus sanguineus* and *Dermacentor variabilis*)-infested and weekly reinfested dogs. The amitraz/imidacloprid combination and the fipronil treatment achieved significantly fewer fleas than the controls. The amitraz collar alone guaranteed significantly fewer ticks (Young and Ryan 1999). The fipronil-treated dogs possessed significantly fewer fleas, and fewer numbers of dogs were infested in that group compared with the amitraz/imidacloprid combination treatment group on days 10, 17, 24, and 31. Also significantly fewer ticks and fewer numbers of tick-infested dogs than in the amitraz/imidacloprid group were recorded on days 10 and 17. The fipronil-treated dogs also showed significantly fewer ticks than the solely amitraz-treated dogs on day 10. Furthermore the fipronil-treated group possessed a higher efficacy against fleas and ticks than the two other treatment groups during the month after treatment (Young and Ryan 1999).

The influence of water in form of washing of the bedding of cats has been tested besides the effect of vacuuming cats' bedding (Ewald-Hamm and Krebber 1998). A form of impregnation of animal bedding through contact with treated pets and thus a larvicidal activity of imidacloprid had been proven (see chapter 'Larvicidal Effect of Imidacloprid and Efficacy Enhancement (with PBO)'). The influence of washing and vacuuming such impregnated bedding was tested to see the effect on the larvicidal action and also to evaluate ways of treatment to remove any chemical compound in animals' bedding (Ewald-Hamm and Krebber 1998). Though the concentration of imidacloprid in the cloths established through contact with treated cats was very low, it was still sufficient to show a significant larvicidal effect when flea eggs were placed on the cloths and incubated under suitable developmental conditions. Cloths were exposed to treated cats for 30 hours (six hours on five consecutive days), cut in half and then analyzed for imidacloprid either without any other processing or they were machine washed and analyzed or vacuumed for a maximum of 30 seconds and consecutively analyzed for imidacloprid. Washing reduced the concentration of imidacloprid in the cloths below the limit of quantitation, which was 0.5 mg/kg fabric corresponding to 0.02 mg/fabric sample. Vacuuming removed approximately 50% of the compound from the cloths. So, if maintenance of larvicidal activity is desired and cleaning of the cats' bedding is routine, washing or vacuuming should be conducted immediately prior to the monthly application of Advantage®. If reduction or removal of imidacloprid in the pets' bedding is required, the compound's concentration will be significantly reduced by normal vacuuming and/or washing at any time (Ewald-Hamm and Krebber 1998).

In field studies testing the efficacy of imidacloprid in flea control on dogs (Hopkins 1997b) three treatments were needed in two of 22 households. At these two sites the dogs swam on a frequent basis. One treatment was needed in twelve and two treatments were needed in eight households to eliminate a long-standing flea problem successfully (Hopkins 1997b).

A similar condition occurred in a field study in Italy (Genchi et al. 2000), in which owners were requested not to bathe their animals during the trial. Nevertheless 224 of

1939 dogs (11.5%) underwent baths or shampoos after imidacloprid treatment. Here, no statistical difference was found comparing flea counts from bathed and unbathed dogs, and the overall efficacy was not affected (Genchi et al. 2000) (Fig. 44). In some cases, dogs were bathed repeatedly (two to four times) throughout the study. The clinical data assessed during the study (regarding itching, alopecia, FAD) tend to show that bathed dogs had a more gradual and less significant relief of signs. FAD was already more severe at the beginning of the study in the group of subjects thereafter submitted to baths or shampoos by the owners. It was presumed that bathing possibly alters the skin environment, worsening the allergic signs from flea bites and delaying the reparative process even in the face of an effective treatment. However, it could also be possible that animals showing more severe signs of dermatitis might have been washed by their owners in the belief that frequent bathing relieves the signs (Genchi et al. 2000).

Even though occasional bathing or water immersion did not significantly compromise the duration of efficacy of imidacloprid, the variability in shampoos having greater degreasing activity or medicinal properties as well as the frequency of bathing are two large factors affecting the duration of activity which cannot be so easily predicted (Arther and MacDonald 1997). According to the same authors a more conservative bathing approach (not more frequently than once weekly) and a reapplication of the parasiticide every three rather than every four weeks is recommended especially during the early phase of initial treatments of patients with FAD, where pyoderma may be present and involve scaling and crusting. The regimen of bathing once weekly combined with a three week interval of application has been satisfactory in controlling fleas (Arther and MacDonald 1997).

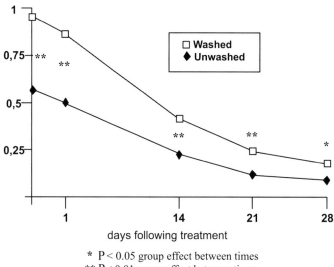

* P < 0.05 group effect between times
** P < 0.01 group effect between times

Fig. 44. Flea allergy dermatitis: comparison between bathed and unbathed dogs

# 'Umbrella Effect'

Safety evaluation in two controlled studies with 27 dogs from eight weeks of age and a treatment with up to 50 mg imidacloprid 10% spot-on/kg b.w. on three consecutive days or for eight consecutive weeks at seven day intervals have not produced any clinical side effects after treatment nor any evidence of dermal irritation at the application site. There was also no indication of any clinically significant trends developing in serum chemistry or hematology parameters and no histologic lesions found in any of the animals that might have been induced by the test article. Body weights in adult dogs maintained or slightly increased, while puppies achieved acceptable weight gains during the study (Shmidl and Arther 1995a, 1995b). For detailed information see also chapter 'Toxicology and Pharmacology of Imidacloprid'. There is only the suggested contraindication that imidacloprid spot-on should not be used in puppies younger than eight or kittens younger than ten weeks of age. The exclusion of unweaned litter in the course of treatment is a condition which is indicated in numerous products, available for the treatment of adolescent and adult animals, but generally excluding the suckling offspring. Alternative methods are therefore required for the control of fleas on suckling animals (Jacobs and Hutchinson 1998). Concentrating on this age category a study was conducted by Jacobs and Hutchinson (1998), testing the control of fleas on kittens via an imidacloprid-treated queen, also proclaimed as 'umbrella effect'. Treatment was applied with imidacloprid according to label instructions. Queens were challenged with flea infestations on a weekly basis beginning with 40 fleas, later increasing to 60 fleas until day 56 p.t.. Forty-eight hours after each infestation, 70-100% of the total number of fleas on the untreated queen and her litter were found to be on the kittens. Considering the fact that the untreated group only consisted of one queen, it cannot be concluded as a typical feature, but may nevertheless provide a tendency. It seemed that fleas may have a preference for kittens, with the number of fleas on the litter independent of the number of kittens in the litter. In other words, it was suggested that the same proportion of the inoculation dose will transfer to the litter however many kittens there may be (Jacobs and Hutchinson 1998). If the number of kittens is independent of the number of fleas transferred, then the litter and not the individual kitten must be regarded as the correct statistical unit (Jacobs and Hutchinson 1998). Relying on these suggestions, even though they have to be qualified because of the small number of test animals (one untreated and two treated queens), the study appeared to show a powerful beneficial effect exerted by the monthly treated queen on flea infestation of the kittens. The flea control maintained >95% for at least three weeks after the first and two weeks after the second treatment (Jacobs and Hutchinson 1998). More than 85% protection of the kittens was maintained throughout the first seven weeks. Declining efficacy in weeks 7 and 8 might have been the consequence of decreasing contact between queens and their litter. If the efficacy is wholly or partly dependent on either a direct transfer of imidacloprid from the coat of the dam to the fur of the kittens, or a continual exchange of fleas between the dam and her litter, then a decreasing contact might influence the results of protection. Even though the results of the study by Jacobs and Hutchinson (1998) have to be interpreted with caution, due to the small number of litters and the unknown

Table 42. Flea efficacy (in %) of imidacloprid on the treated queens and protection of the untreated kittens in a study with four weeks treatment interval

| | | | | | Weeks p.t. | | | | | |
|---|---|---|---|---|---|---|---|---|---|---|
| | 0^ | 1 | 2 | 3 | 4† | 4# | 5 | 6 | 7 | 8 |
| Kittens | 100 | 100 | 100 | 98 | 81 | 100 | 99 | 95 | 87 | 74 |
| Kittens and Queens | 100 | 100 | 100 | 97 | 81 | 100 | 98 | 96 | 84 | 69 |

^ = comparison of pre- and post-treatment counts; all others comparison of treated and control
† = pre-treated
# = post-treated

validity of the assumptions made in the calculation (i.e. regarding the dynamics of flea transfer between queen and kittens), the apparent efficacy as shown in Table 42 present a strong tendency that there will be a protection of the kittens via the treated queen (Jacobs and Hutchinson 1998).

## Small Domestic Animals

### Rabbits

The rabbit flea, *Spilopsyllus cuniculi* (Dale 1878), has been reported to infest wild as well as pet rabbits (Rothschild 1965; Pinter 1999; Vaughan and Coombs 1979). In contrast to the cat flea's biology (see chapter 'Developmental Cycle of Fleas'), the biology of *S. cuniculi* is closely adapted to its host reproductive cycle (reviewed by Marshall 1981). Most of the time during the rabbit reproductive cycle *S. cuniculi* feeding on the host remain sexually immature. The sexual development is not completed until the infested doe approaches the end of pregnancy, when the fleas congregate on and attach to the ears of their hosts. Mating finally does not take place until after parturition and will then only happen if there are live kittens present. Over the next week, females periodically leave the host (this also in strong contrast to *C. felis*) to lay their eggs in the nesting materials. The new generation of adult fleas appears in two waves. The first leave their cocoons spontaneously after 15-30 days at the time when the doe and her litter are about to leave the nest, while the second wave is delayed until visiting animals subsequently stimulate emergence (Sobey et al. 1974). Besides *S. cuniculi* which is the vector for myxomatosis and the major species on the European rabbit, *Oryctolagus cuniculus*, *Cediopsylla simplex* (the common eastern rabbit flea) and *Odontopsyllus multispinus* (the giant eastern rabbit flea) may also infest rabbits, the latter ones e.g. the North America cotton tailed rabbit (*Sylvilagus floridanus*) (Timm 1988). But infestations of pet rabbits with one of these flea species are only likely to occur where there is direct or indirect contact with wild rabbits (Jacobs 2001). Pet rabbits are, however, often exposed to the almost ubiquitous flea species of the domestic environment, the cat flea, *C. felis*. They are vulnerable to attack by host-seeking cat fleas (Jacobs 2001). Veterinary clinics worldwide are well aware of cat flea infestation in rabbits (Fukase et al. 2000). Concerning an effective and especially safe product for flea control in this animal species, little information is available on the efficacy of topical flea products and their use, due to the fact that many of them are licensed for cats and dogs, but none specifically for rabbits. Caution is needed when extrapolating from recommendations designed for the cat or dog, as rabbits obviously differ with regard to skin and hair-type. Furthermore differences in medicinal product distribution and metabolism can also potentially influence efficacy and safety (Jacobs 2001).

Two carefully controlled experimental studies to measure the efficacy and duration of efficacy of imidacloprid 10% spot-on against *C. felis* on artificially infested rabbits were conducted at the Bayer laboratories in Monheim, Germany, and at the Royal Veterinary College, London, England (Jacobs et al. 2001; Hutchinson et al. 2001a). Each used two groups of six individually caged rabbits. Allocation to groups was made by random, or the rabbits were first ranked according to their susceptibility to fleas. This was ascertained by counting the fleas that established on each rabbit after a uniform pre-trial (day -8) infestation. Rabbits in the treatment groups were given a single spot-on treatment on day 0. As all the rabbits weighed less than 4 kg, each received 0.4 ml of the 10% topical formulation (Advantage® 40 for Cats) applied directly onto the skin of the neck just behind the base of the skull. This provided a minimum dose of 10 mg/kg. In the British study, fleas were placed on the rabbits on days -8, -1, 7, 14, 21 and 28. An infestation of 100 *C. felis* gave a consistent and adequate level of infestation without excessive discomfort to the rabbit. Live fleas were observed on the floor of the cages of the untreated controls. This contrasts sharply with experience in the same laboratory with *C. felis* infestations on cats, as in this case the fleas are almost invariably found only on the host. Similar infestations on rabbits in Germany resulted in much lower establishment rates and so two infestations with 100 fleas were given on separate days prior to each observation point (i.e. on days -3 and -1 as well as days 4 and 7). It is not known if the difference between establishment rates in the British and German studies reflects the strain of cat flea used, the breed of rabbit or other unknown factors.

To measure curative efficacy against an established *C. felis* population, flea counts were performed eight and 24 hours after treatment in the British trial, and after 24 hours in the German study. Flea burdens were reduced by 96% within eight hours of treatment and 100% efficacy was recorded at 24 hours. For evaluation of residual protective activity, counts were made 24 hours after each subsequent infestation. In the British study, flea counts following reinfestation one, two, three and four weeks p.t. were reduced by 95, 81, 79 and 68%, respectively (Fig. 45). These differences in flea counts between treated and control groups are statistically significant (P <0.002). The German results differed in that no protection against reinfestation was apparent one week after treatment.

The third efficacy study was a clinical field trial conducted at the Meiji Pharmaceutical University, Tokyo, Japan (Fukase et al. 2000) in which 30 naturally infested rabbits presented to a veterinary clinic were treated with imidacloprid 10% spot-on at 10 mg/kg b.w. were compared with 30 untreated controls. Infestations were mostly *C. felis* but a few additional dog fleas (*C. canis*) were noted on two of the rabbits. Initial flea burdens numbered five to ten on most animals but some had between ten and 20. Flea

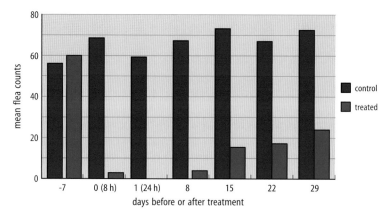

**Fig. 45.** Mean flea counts of a group of six rabbits treated with imidacloprid 10% spot-on on day 0 and a similar group kept as untreated controls

counts were performed day 1, 2, 3, 5 and 7 p.t. and treated animals were also combed after two and four weeks. Each rabbit returned to its owner's home after each visit to the clinic. With the exception of one animal on one occasion, fleas were found on all of the controls at every count (Table 43).

This confirms that the rabbits were continually exposed to reinfestation in their home environment. In contrast, only three of 30 treated rabbits harbored fleas one day p.t., and all were free from fleas between days 2 and 7. Small numbers of fleas (<5) were found on one rabbit two weeks p.t. and on three rabbits at four weeks. Thus, the Japanese clinical observations closely mirror the pattern of results seen in the British experimental study with substantial, although incomplete, protection against reinfestation persisting for at least one month.

Ancillary safety studies were conducted in Japan (Fukase et al. 2000) and Germany (Andrews 1999, personal communication). In the Japanese study, groups of six laboratory rabbits, weighing 1.2-1.5 kg, were kept either as untreated controls or dosed at 10 mg/kg, 100 mg/kg or 10 mg/kg b.w. daily for three consecutive days. No clinical abnormalities were detected and no significant differences were found between groups when hematological and biochemical parameters were compared. Similarly, in German dermal toxicity tests, with 1-, 3- and 5-times the recommended minimum dosage failed to

**Table 43.** Number of rabbits infested with fleas following imidacloprid 10% spot-on treatment at day 0 of the study

| Test group | No. of rabbits | Dosage (mg/kg b.w.) | No. of rabbits infested on days p. t. | | | | | | | |
|---|---|---|---|---|---|---|---|---|---|---|
| | | | $0^1$ | 1 | 2 | 3 | 5 | 7 | 14 | 28 |
| Untreated | 30 | 0 | 30 | 30 | 30 | 30 | 30 | 29 | —² | —² |
| Treated | 30 | 10 | 30 | 3 | 0 | 0 | 0 | 0 | 1 | 3 |

¹ = before treatment; ² = not examined

induce any observable abnormality in pairs of rabbits with body weights of 2.5-2.9 kg. No ill-effect resulted when 0.1 ml of the 10% imidacloprid spot-on formulation was administered orally to a rabbit to mimic possible intake by licking and grooming after topical application.

These experimental and clinical field trials demonstrate that imidacloprid 10% spot-on provides excellent efficacy against a resident flea infestation on rabbits. The subsequent residual protective effect is not as complete as that seen on treated cats or dogs but is nevertheless substantial, persisting for four weeks in two of the three trials. Results from the third study suggest that this duration of action may be variable, but determining factors have yet to be identified. No adverse effects were seen in any treated animal. The safety of topically applied imidacloprid was confirmed by experimental studies in which rabbits were deliberately overdosed without ill-effect.

## Ferrets

There are few medicines with indications for ferrets, despite the popularity for keeping ferrets especially in the USA and Europe. Therapy for fleas on ferrets could seem unnecessary as there is no species of 'ferret flea', though it is recognized that ferrets may occasionally become infested with fleas of their prey. However, particularly now ferrets are kept as household pets, the ubiquitous cat flea, *C. felis* has adopted the ferret as an additional host species.

Flea infestation in ferrets can produce the disease entities normally associated with flea infestation in dogs and cats. For example, flea-infested ferrets may develop signs of hypersensitivity, with lesions similar to those seen in cats: papulocrustous dermatitis and self-inflicted alopecia. It is unknown how satisfactorily ferrets host cat fleas, though it is clear from the studies described here that individual cat fleas can survive well on ferrets.

There does not appear to be much guidance in the literature about treating ferrets with fleas. A letter to The Veterinary Record (Oxenham 1996) reported the successful treatment of 50 ferrets with diazinon 20% w/v (Droplix®, Virbac) at dose rates of 0.1 ml for ferrets 600-1500 g b.w. and 0.15 ml over 1500 g b.w., a treatment no longer available. Text books tend to recommend treating ferrets, in-contact ferrets and the environment as for flea infestation in cats (Fox 1998). Fox (1998) speculated that topical imidacloprid might be useful for controlling fleas on ferrets, following studies showing successful control of fleas in cats and dogs.

Two studies have recently been conducted to examine the safety and efficacy of imidacloprid 10% spot-on against the cat flea, *C. felis,* on European ferrets, *Mustela putorius furo*. Imidacloprid was either applied at a dose rate of 10 mg/kg b.w.(Jacobs, Royal Veterinary College, London, UK) or as one 0.4 ml pipette applied to each animal (McCall, TRS Labs., Athens, GA, USA)(Fisher et al. 2001).

The ferrets were experimentally infested with 60 unfed adult *C. felis* on the day before allocation and the day before treatment, followed by once weekly infestations of 60 adult fleas from day 7 to day 28. Flea counts were performed prior to allocation, then eight and 24 hours p.t.. Subsequent counts were carried out 24 hours after each weekly reinfestation. Fleas removed by combing were not returned to the ferret. In both studies, efficacy of treatment at each time point was calculated by comparing the geometric mean flea counts of the treated and control groups. In both studies, no abnormalities were observed after treatment and all ferrets remained normal throughout. Recovery of live fleas from the untreated ferrets was remarkable compared to recovery in cats, for example. In the study conducted at the Royal Veterinary College, London, between 77 and 92% of fleas were counted on each occasion (Hutchinson et al. 2001b) and in the study conducted at TRS Labs., Athens, between 87.8 and 100% of fleas were accounted for.

Treated male ferret body weights ranged from 0.85-1.1 kg (USA) and 1.41-1.61 kg (UK). The smaller female ferrets weighed between 0.6-0.7 kg (USA) and 0.79-1.05 kg (UK). The actual dosage ranged from 36.4-47.05 mg/kg for males and 57.1-66.6 mg/kg for females in the study where a single pipette containing 0.4 ml of the imidacloprid 10% spot-on solution was applied.

Percentage efficacies calculated from the geometric flea counts are shown in Fig. 46 and Fig. 47. Efficacy against a flea infestation present at the time of treatment, assessed in the UK study was >95% by eight hours, and 100% by 24 hours after treatment. One week after treatment efficacy against a challenge infestation applied 24 hours previously remained >90% in both studies. However, thereafter efficacy declined, most rapidly in the UK study where day 15 efficacy was 55%. At the end of the study, four weeks p.t., residual efficacy in the US study was 72%.

The studies demonstrated that imidacloprid was well tolerated when administered at a dose rate of 10 mg/kg b.w. or one 0.4 ml pipette per animal. Both dose rates were highly effective at killing an established flea burden and retained good activity for a period of one week after treatment, when good efficacy is defined as 90% or more.

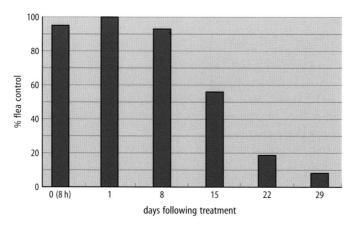

Fig. 46. Percentage efficacy of imidacloprid 10% spot-on administered to ferrets at a dosage of 10 mg/kg b.w. against *C. felis*

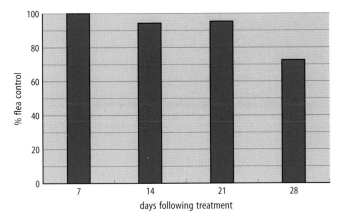

Fig. 47. Percentage efficacy of imidacloprid 10% spot-on administered to ferrets at a dose-rate of one 0.4 ml pipette per ferret against *C. felis*

Using one 0.4 ml pipette per animal the duration of good efficacy was extended up to three weeks after treatment. Duration of activity therefore appears to be dose-related, with increased dose rate extending the useful activity.

## Rodents and Other Small Animals

As described before for pet rabbits other small animals kept in households with or without cats and dogs are targets for flea infestation. Numerous anecdotal information is available where especially rats as well as mice kept as pet animals have been presented to veterinary clinics with flea infestation. While there is no record on identification of the flea genus and species, one can assume that cat fleas might be the cause of most of these consultations. Imidacloprid 10% spot-on has been used successfully to eliminate the existing flea burden (Bayer AG Germany research data on file). The clinical observations are clear to confirm the therapeutic effect of imidacloprid, while no information can be given on the duration of efficacy and the effect on reinfestations.

## Ectoparasitic Spectrum of Activity

### Louse Infestation

Even though ticks did not prove to be sensitive in form of a complete control by imidacloprid (Cruthers et al. 1999; Young and Ryan 1999 (see chapter ' Environmental and Habitual Factors Influencing Imidacloprid Treatment')), the spectrum of activity of imidacloprid in the field of ectoparasites is not confined to fleas. Imidacloprid in the dosage as recommended for flea control has also proven to be highly effective against both sucking lice (*Linognathus setosus* Olfers 1816) and biting or chewing lice (*Trichodectes canis* DeGeer 1778) (Fig. 48), over a 6-week period of a study (Hanssen et al. 1999). Lice are reported to be the most common ectoparasites in dogs in regions of the northern hemisphere (Persson 1973; Christensson et al. 1998), outnumbering fleas and ticks, in regions where fleas are often rare or even nonexistent. In the southern hemisphere, they are at least concurrent parasites on dogs kept under robust, outdoor housing conditions.

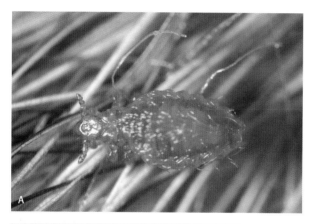

Fig. 48. A,B. Picture of biting *(Linognathus setosus)* (**A**) and sucking *(Trichodectes canis)* (**B**) lice from dogs (original size ~ 2mm)

Most products require two treatments at an interval of one week (Bowman 1999) to control lice emerging from eggs that have survived the first treatment. Fipronil as a spray formulation was suggested to be sufficient in a single treatment against lice on dogs (Cooper and Penaliggon 1996). In naturally infested dogs with sucking or biting lice a single treatment with imidacloprid 10% spot-on was highly effective in lice control over six weeks study period. Examinations 24 hours p.t. as well as two, four, and six weeks after treatment did not record any lice from any of 27 dogs, demonstrating 100% efficacy. The remission of the generalized parasitic dermatitis present in all 27 dogs, as well as signs of alopecia, was first seen at two weeks after treatment. The remission of pruritus was reported to start between three and 14 days after treatment. Interestingly, pruritus was seen to be present even in the absence of an active louse infestation for up to two to four weeks after treatment (Hanssen et al. 1999). In some of the treated dogs (nine) concurrent therapy with corticosteroids and antibiotics was not excluded, due to the clinical status of the skin presented at different examination times. Duration of efficacy exceeded the egg-to-egg development period of *L. setosus* (eight to ten days (Zlotorzycka et al. 1974)) and *T. canis* (three to five weeks (Zlotorzycka et al. 1974)). The residual activity was sufficient to affect the next generation of lice, so that one treatment proved to be enough for complete control of lice on dogs (Hanssen et al. 1999).

## Sheep Ked Infestation

The effects of imidacloprid on sheep ked (*Melophagus ovinus* Linné 1758) infestation on sheep was studied using sheep from a research farm (Mehlhorn et al. 2001a). Sheep keds of the species *Melophagus ovinus* are wingless parasitic insects on sheep belonging to the dipteran family Hippobiscidae (Fig. 49). They are transmitted constantly by body contact with sheep of a herd after one animal has become infested at a resting site. The whole life cycle takes place on the host. About three weeks after copulation the female keds lay brownish, 3x4 mm long puparia which are attached by a glue-like substance to the sheep´s hair. The preadult stage hatches five to ten days later from the puparian cocoon and starts sucking blood. Even if sheep keds are permanent parasites that do not leave the host, the puparia are not so well fixed and therefore might be found at the sheep resting site and act as reservoir for transmission of infection. Besides skin infections with loss in wool quality and meat production as their major clinical impact, sheep keds are known to transmit diseases such as trypanosomiasis.

In the research study reported by Mehlhorn et al. (2001a) naturally infested sheep were treated with imidacloprid at a dosage of 10 mg/kg b.w. and the efficacy on sheep as well as the in vitro effects on the keds were recorded. Imidacloprid acted rapidly on all motile stages of sheep keds. Within three to four minutes after exposure the keds became immobile and their legs and the abdomen started tetanic trembling movements for 15 to 30 minutes, leading to death. On filter paper containing imidacloprid sheep keds became affected as shown by electron microscopy demonstrating lethal destruction

**A**                                                   **B**

**Fig. 49. A, B.** Sheep ked (*Melophagus ovinus* Linné 1758). Macroscopic pictures of the adult (**A**) and the puparium (**B**) (original size of adult 10 mm)

of the ganglia, nerve cords and related muscle fibers. On sheep imidacloprid was available and active for more than one month within the wool. Body contact between treated mothers and their untreated lambs gave protection from infestation with these ectoparasites.

The broad insecticidal activity was thus also verified by this study.

# References

Abbink J (1991) Zur Biochemie von Imidacloprid (The biochemistry of imidacloprid). Pflanzenschutz-Nachr Bayer 44:183-194

Adams WH, Emmons RW, Brooks JE (1970) The changing ecology of murine (endemic) typhus in southern California. Am J Trop Med Hyg 19:311-318

Akin DE (1984) Relationship between feeding and reproduction in the cat flea *Ctenocephalides felis* (Bouché) (Siphonaptera: Pulicidae). MS Thesis, University of Florida, Gainesville

Albuquerque EX, Eldfrawi AT, Eldefrawi ME (1979) The use of snake toxins for the study of the acetylcholine receptor and its ion-conductance modulator. In: Lee CY (ed) Snake venoms. Springer-Verlag, Berlin, pp 377-402

Allen SK, McKeever PJ (1974) Skin biopsy techniques. Vet Clin North Am 4:269-280

Amin OM (1966) The fleas (Siphonaptera) of Egypt: distribution and seasonal dynamics of fleas infesting dogs in the Nile valley and delta. J Med Entomol 3:293-298

Amin OM (1976) Host associations and seasonal occurrence of fleas from southeastern Wisconsin mammals, with observations on morphologic variations. J Med Entomol 13:179-192

Andrews P (1996a) Bayer internal report, 19 Feb., 1996, AH-D - ID 15851

Andrews P (1996b) Bayer internal report, 21 Feb., 1996, AH-D - ID 15852

Andrews P, Bomann W (1996) Bayer internal report, 1 April, 1996, AH-D - ID 16077

Arther RG, MacDonald JM (1997) Clinical and research perspectives. Suppl Comp Cont Educ Pract Vet 19 (5):43-46

Arther RG, Cunningham J, Everett R (1997a) Evaluating the effects of shampooing or repeated water exposure on the residual efficacy of Advantage® (imidacloprid) for flea control on dogs. In: Proc 4th Int Symp Ectoparas Pets, Riverside, California, April 1997, pp 73-74

Arther RG, Cunningham J, Dorn H, Everett R, Herr LG, Hopkins T (1997b) Efficacy of imidacloprid for removal and control of fleas (*Ctenocephalides felis*) on dogs. Am J Vet Res 58:848-850

Azad AF (1990) Epidemiology of murine typhus. Ann Rev Entomol 35:535-569

Bacigalupo J (1931) Evolution de l'*Hymenolepis fraterna* Stiles, chez *Pulex irritans* L., *Xenopsylla cheopsis* Rothschild et *Ctenocephalides canis* Curtis. Ann Parasitol Hum Comp 9:339-343

Bacot A (1914) A study of the bionomics of the common rat fleas and other species associated with human habitation, with special reference to the influence of temperature and humidity of various periods in the life history of the insects. J Hygiene 13 (Plague Suppl 3):447-654

Bacot AW, Ridewood WG (1915) Observations on the larva of fleas. Parasitol 7:157-175

Bai D, Lummis SCR, Leicht W, Breer H, Sattelle DB (1991) Actions of imidacloprid and a related nitromethylene on cholinergic receptors of an identified insect motor neuron. Pestic Sci 33:197-204

Baker KP (1971) Intradermal tests as an aid to the diagnosis of skin disease in dogs. J Small Anim Pract 12:445-452

Baker KP (1974) Observations on allergic reactions in arthropod parasites. Irish Vet J 28:65-70

Baker KP (1977) The hypersensitive response of the skin to fleas with observations on treatment and control. Irish Vet J 31:141-147

Baker KP, Elharam S (1992) The biology of *Ctenocephalides canis* in Ireland. Vet Parasitol 45:141-146

Baker KP, Hatch C (1972) The species of fleas found on Dublin dogs. Vet Rec 91:151-152

Baker KP, Mulcahy R (1986) Fleas on hedgehogs and dogs in the Dublin area. Vet Rec 119:16-17

Baker KP, O'Flanagan J (1975) Hypersensitivity of dog skin to fleas - a clinical report. J Small Anim Pract 16:317-327

Baker KP, Thomsett LR (1990) Canine and feline dermatology. Blackwell Scientific, Oxford, pp 161-165

Baker N (1984) Musing the relationship between a dog and its fleas. Vet Med 79:1037-1039

Baker N (1985) The touch-and-go relationship of a dog and its fleas. Vet Med 80 (Suppl):6-7

Bardt D (1996) Bayer internal report, 15 Oct., 1996, AH-D - ID 17057

Bardt D, Schein E (1996) Zur Problematik von therapieresistenten Flohpopulationen am Beispiel des Stammes "Cottontail". Kleintierpraxis 41:561-566

Barnes EH, Dobson RJ, Barger IA (1995) Worm control and anthelmintic resistance: adventures with a model. Parasitol Today 11:56-63

Beaucournu JC (1975) *Ctenocephalides brygooi* sp. n. (Siphonaptera, Pulicidae), parasite du Viverridé endémique malgache *Fossa fossana* (Muller). (*Ctenocephalides brygooi* n. sp. (Siphonaptera, Pulicidae), a parasite of Malagasy endemic Viverrid *Fossa fossana* (Muller)). Bull Soc Pathol Ex 67:435-443

Beaucournu JC (1990) Les puces synanthropes. Bull Soc Franç Parasitol 8:145-156

Beck AM (1973) The ecology of stray dogs. York, Baltimore

Bell G (1997) Introducing Advantage - The last word in flea control. Irish Vet J 50:375, 377-379

Benjamini E, Feingold BF, Kartman L (1961) Skin reactivity in guinea pigs sensitized to flea bites: The sequence of reactions. Proc Soc Exp Biol Med 108:700-702

Benjamini E, Feingold BF, Young JD, Kartmann L, Shimizu M (1963) Allergy to flea bites. IV. In vitro collection and antigenic properties of the oral secretion of the cat flea, *Ctenocephalides felis felis* (Bouché). Exp Parasitol 13:143-154

Bennett GW, RK Lund (1977) Evaluation of encapsulated pyrethrins (Sectrol™) for German cockroach and cat flea control. Pest Control 45:44, 46, 48-50

Benson JA (1989) Insect nicotinic acetylcholine receptors as the targets for insecticides. BCPC Mono. No. 43 Progress and prospects in insect control.

Benton AH, Lee SY (1965) Sensory reactions of Siphonaptera in relation to host-finding. Am Midl Nat 74:119-125

Beresford-Jones WP (1974) The fleas *Ctenocephalides felis felis* (Bouché, 1833), *Ctenocephalides canis* (Curtis, 1826), and the mite *Cheyletiella* (Canestrini, 1886) in the dog and cat: their transmissibility to humans. In: Soulsby EJL (ed) Parasitic zoonoses, clinical and experimental studies. Academic Press, London, pp 383-390

Beresford-Jones WP (1981) Prevalence of fleas on dogs and cats in an area of central London. J Small Anim Pract 22:27-29

Bevier-Tournay DE (1989) Fleas and flea control. Curr Vet Therapy 10:586-592

Blackmon DM, Nolan MP (1984) *Ctenocephalides felis* infestation in Holstein calves. Agric Pract 5:6-8

Blagburn BL, Hendrix CM (1989) Systemic flea therapy: an overview of flea biology and control. In: Perspectives in systemic flea control. Publication 2075, College of Veterinary Medicine, Auburn University, Auburn, pp 4-9

Blagburn BL, Lindsay DS (1995) Ectoparasiticides. In: Adams HR (ed) Veterinary pharmacology and therapeutics. Iowa State University Press, Ames, pp 984-1003

Blagburn BL, Bach T, Bledsoe DL, Denholm I, Dryden MW, Hansen O, Hinkle NC, Hopkins T, Jacobs DE, Mehlhorn H, Mencke N, Payne PA, Rust MK, Vaughn MB (2001) Advantage® international flea susceptibility monitoring initiative – a 2001 update. Suppl Comp Cont Educ Vet Pract 23 (4A):4-7

Blagburn BL, Hendrix CM, Vaughan JL, Lindsay DS, Barnett SH (1995) Efficacy of lufenuron against developmental stages of fleas (*Ctenocephalides felis felis*) in dogs housed in simulated home environments. Am J Vet Res 56:464-467

Blagburn BL, Vaughan JL, Lindsay DS, Tebbitt GL (1994) Efficacy dosage titration of lufenuron against developmental stages of fleas (*Ctenocephalides felis felis*) in cats. Am J Vet Res 55:98-101

Blanc G, Baltazard M (1914) Revue chronologique sur la transmission et la conversation naturelles des typhus. III. Comportement des virus de la fievre boutonneuse et de la fievre pourpree chez les puces *Xenopsylla cheopsis* et *Ctenocephalides canis*. Arch Inst Pasteur Maroc 2:602-610

Borror DJ, DeLong DW, Triplehorn CA (eds) (1981) Order Siphonaptera. In: An introduction to the study of insects, 5th edn. WB Saunders Co, Philadelphia, pp 620-628

Bossard RL (1997) Evaluation and use of bioassays for surveying insecticide susceptibility of cat fleas, *Ctenocephalides felis felis* (Bouché), in relation to resistance. Ph.D. Dissertation, Kansas State University, Manhattan

Bossard RL, Hinkle NC, Rust MK (1998) Review of insecticide resistance in cat fleas (Siphonaptera: Pulicidae). J Med Entomol 35:415-422

Bossard R, Dryden M, Broce A (1996) Insecticide resistance of cat fleas, *Ctenocephalides felis felis*, in the United States. In: Proc 41st Ann Meet. Am Ass. Vet Parasitol, Louisville, Kentucky, July 1996, No. 55, p 48

Boucias DG (1996) Imidacloprid - chemical synergist for microbal control agents of termites. Abstr Pap Am Chem Soc 212 Meet (Pt.1) AGRO 018:1996

Bourdeau P (1983) La dermatite par «allergie» aux piqûres des puces chez le chien (D.A.P.P.) 1re partie. Clinique, épidémiologie et pathogénie. (Flea allergy dermatitis in dogs. 1. Clinical signs, epidemiology and pathogenesis.) Le Point Vet 15:17-25

Bourdeau P, Blumenstein P (1995) Flea infestation and *Ctenocephalides* in the dog and cat; parasitological, biological and immunological aspects in the west of France. Ann Congr Europ Soc Vet Dermat, Barcelona (Abstr.) pp 1-6

Bowman DD (ed) (1999) Orders Anoplura, bloodsucking lice, and Mallophaga, chewing lice. In: Georgis' parasitology for veterinarians, 7th edn. WB Saunders, Philadelphia, pp 29-37

Breer H, Sattelle DB (1987) Molecular properties and functions of insect acetylcholine receptors. J Insect Physiol 33:771-790

Briggs OM (1986) Flea control on pets in Southern Africa. J South Afr Vet Assoc 57:43-47

Brown AWA (1958) The spread of insecticide resistance in pest species. Adv Pest Cont Res 2:351-413

Brown AWA, Pal R (1971) Insecticide resistance in arthropods, 2nd edn. WHO, Geneva

Brown JH, Roughgarden J (1989) US ecologists address global change. Trends Ecol Evol 4:255-256

Bruce WW (1948) Studies on the biological requirements of the cat flea. Ann Entomol Soc Am 41:346-352

Byron DW (1987) Aspects of the biology, behaviour, bionomics, and control of immature stages of the cat flea *Ctenocephalides felis felis* (Bouché) in the domiciliary environment. Ph.D. Dissertation, Virginia Polytechnic Institute and State University, Blacksburg

Byron DW, Robinson WH (1986) Research on household flea control. Pest Manage 5:29-31

Cachelin AB, Jaggi RJ (1991) Beta subunits determine the time course of desensitization in rat alpha 3 neuronal nicotinic acetylcholine receptors. Pflügers Arch 419:579-582

Cahill M, Gorman K, Day S, Denholm I, Elbert A, Nauen R (1996) Baseline determination and detection of resistance to imidacloprid in *Bemisia tabaci* (Homoptera: Aleyrodidae). Bull Entomol Res 86:343-349

Carlotti DN, Jacobs DE (2000) Therapy, control and prevention of flea allergy dermatitis in dogs and cats. Review article. Vet Dermatol 11:83-98

Chamberlain KW, Baker E (1974) Diagnostic methods in allergic disease. Vet Clin North Am 4:47-56

Chen HT (1933) On a method of expelling disintegrated tapeworms in *Ctenocephalides felis*. Lingnan Sci J 12:43-48

Chen HT (1934) Reactions of *Ctenocephalides felis* to *Dipylidium caninum*. Z Parasitenkd 6:603-637

Chesney CJ (1995) Species of flea found on cats and dogs in south west England: further evidence of their polyxenous state and implications for flea control. Vet Rec 136:356-358

Cheung H, Clarke BS, Beadle DJ (1992) A patch-clamp study of the action of a nitromethylene heterocycle insecticide on cockroach neurones growing in vitro. Pestic Sci 34:187-193

Christensson D, Zakrisson G, Holm B, Gunnarsson L (1998) Lus hos hund i Sverige. Svensk Veterinartidn 50:189-191

Coetzee BB, Stanford GD, Davis DAT (1987) Resistance by the blue tick (*Boophilus decoloratus*) to the synthetic pyrethroid, fenvalerate. Onderstepoort J Vet Res 54:83-86

Collart MG, Hink WF (1986) Development of resistance to malathion in cat flea (Siphonaptera: Pulicidae). J Econ Entomol 79:1570-1572

Conniff R (1995) When it comes to pesky flea, ignorance is bliss. Smithsonian 26:76-85

Cooper PR, Penaliggon J (1996) Use of fipronil to eliminate recurrent infestation by *Trichodectes canis* in a pack of bloodhounds. Vet Rec 139:95

Corbett JR, Wright K, Baillie AC (eds) (1984) Insecticides acting elsewhere in the nervous system. II. Compounds that bind to the acetylcholine receptor. In: The biochemical mode of action of pesticides, 2nd edn. Academic Press, London, pp 159-162

Cornack KM, O'Rourke PK (1991) Parasites of sheep dogs in the Charleville district, Queensland. Aust Vet J 68:149

Coward PS (1991) Fleas in southern England. Vet Rec 129:272

Crum GE, Knapp FW, White GM (1974) Response of the cat flea, *Ctenocephalides felis* (Bouché), and the oriental rat flea, *Xenopsylla cheopis* (Rothschild), to electromagnetic radiation in the 300-700 nanometer range. J Med Entomol 11:88-94

Cruthers L, Bock E (1997) Evaluation of how quickly imidacloprid kills fleas on dogs. Suppl Comp Cont Educ Pract Vet 19 (5):27

Cruthers L, Guerrero J, Robertson-Plouch C (1999) Evaluation of the speed of kill of fleas and ticks with fipronil or imidacloprid. In: Proc 5th Int Symp Ectoparas Pets, Fort Collins, Colorado, April 1999, No. 27

Cunningham J (1995) Bayer internal report, 25 April, 1995, AH-D - ID 15377

Cunningham J, Everett R (1994) Bayer internal report, 17 Oct. 1995, AH-D - ID 14909

Cunningham J, Everett R, Arther RG (1997a) Efficacy of imidacloprid on large dogs. Suppl Comp Cont Educ Pract Vet 19 (5):28

Cunningham J, Everett R, Arther RG (1997b) Effects of shampooing or water exposure on the initial and residual efficacy of imidacloprid. Suppl Comp Cont Educ Pract Vet 19 (5):29-30

Cunningham J, Everett R, Ryan WG (1999) Assessment of Frontline® Top Spot™ & Advantage® in controlling fleas on dogs shampooed pre-treatment or water immersed after treatment. In: Proc 5th Int Symp Ectoparas Pets, Fort Collins, Colorado, April 1999, No. 23

Davidson DL (1992) 'Superfleas' bounce back? Vet Rec 131:223

Deboer DJ (1989) Survey of intradermal skin testing practices in North America. J Am Vet Med Assoc 195:1357-1363

Deoras PJ, Prasad RS (1967) Feeding mechanisms of Indian fleas X. cheopsis (Roths) and X. astia (Roths). Ind J Med Res 55:1041-1050

Devillers-Thiéry A, Galzi JL, Eiselé JL, Bertrand S, Bertrand D, Changeux JP (1993) Functional architecture of the nicotinic acetylcholine receptor: A prototype of ligand-gated ion channels. J Membr Biol 136:97-112

Devine GJ, Harling ZK, Scarr AW, Devonshire AL (1996) Lethal and sublethal effects of imidacloprid on nicotine-tolerant *Myzus nicotianae* and *Myzus persicae*. Pestic Sci 48:57-62

Devonshire AL, Field LM (1991) Gene amplification and insecticide resistance. Ann Rev Entomol 36:1-23

Dewar AM, Read LA (1990) Evaluation of an insecticidal seed treatment, imidacloprid, for controlling aphids on sugar beet. In: Proc Brighton Crop Protect Conf - Pests and Diseases, pp 731-736

Diehr HJ, Gallenkamp B, Jelich K, Lantzsch R, Shiokawa K (1991) Synthese und chemisch-physikalische Eigenschaften des Insektizids Imidacloprid (NTN 33893) (Synthesis and chemical-physical properties of the insecticide imidacloprid (NTN 33893)). Pflanzenschutz-Nachr. Bayer 44:107-112

Dobri L (1996) Personal communication, cited in Nauen et al. 1998b

Donnellan JF, Clarke BS, Harris R, Cattell KJ (1979) Acetylcholine-binding proteins in the insect CNS as targets for insecticides. In: Insect Neurobiology and Pesticide Action. SCI, London, pp 129-136

Dryden MW (1988) Evaluation of certain parameters in the bionomics of *Ctenocephalides felis felis* (Bouché 1835). MS Thesis, Purdue University, West Lafayette

Dryden MW (1989a) Biology of the cat flea, *Ctenocephalides felis felis*. Comp Anim Pract 19:23-27

Dryden MW (1989b) Host association, on-host longevity and egg production of *Ctenocephalides felis felis*. Vet Parasitol 34:117-122

Dryden MW (1990) Blood consumption and feeding behaviour of the cat flea, *Ctenocephalides felis felis* (Bouché 1835). PhD. Dissertation, Purdue University, West Lafayette

Dryden MW (1993) Biology of fleas of dogs and cats. Comp Cont Educ Pract Vet 15 (4):569-579

Dryden MW (1996) A look at the latest developments in flea biology and control. Vet Med Suppl 3: 3-8

Dryden MW, Blakemore JC (1989) A review of flea allergy dermatitis in the dog and cat. Comp Anim Pract 19:10-17

Dryden MW, Broce AB (1993) Development of a trap for collecting newly emerged *Ctenocephalides felis* (Siphonaptera: Pulicidae) in homes. J Med Entomol 30:901-906

Dryden MW, Broce AB (2000) Susceptibility of fleas to control agents: development of a monitoring system. Suppl Comp Cont Educ Pract Vet 22 (4A):21-22

Dryden MW, Gaafar SM (1991) Blood consumption by the cat flea, *Ctenocephalides felis* (Siphonaptera: Pulicidae). J Med Entomol 28:394-400

Dryden MW, Prestwood AK (1993) Successful flea control. Comp Cont Educ Pract Vet 15:821-831

Dryden MW, Reid B (1993) Investigations of cat flea pupation, cocoon formation and the impact of pupae on a flea control program. In: Knapp FW, Potter M, Labore D (eds) Proc Int Symp Ectoparas Pets, Lexington University Ky Press, Lexington, pp 27-29

Dryden MW, Reid BL (1996) Insecticide susceptibility of cat flea (Siphonaptera: Pulicidae) pupae. J Econ Entomol 89:421-427

Dryden MW, Rust MK (1994) The cat flea: biology, ecology and control. Vet Parasitol 52:1-19

Dryden MW, Smith V (1994) Cat flea (Siphonaptera: Pulicidae) cocoon formation and development of naked flea pupae. J Med Entomol 31:272-277

Dryden MW, Boyer J, Smith V (1994) Techniques for estimating on-animal populations of *Ctenocephalides felis* (Siphonaptera: Pulicidae). J Med Entomol 31:631-634

Dryden MW, Broce BA, Moore WE (1993) Severe flea infestation on dairy calves. J Am Vet Med Assoc 203:1448-1452

Dryden MW, Denenberg TM, Bunch S (2000) Control of fleas on naturally infested dogs and cats in private residences with topical spot application of fipronil and imidacloprid. Vet Parasitol 93:69-75

Dryden MW, Maggid T, Bunch S (1999a) Control of fleas on naturally infested dogs and cats and in private residences with topical spot applications of 9.1% imidacloprid or 9.7% fipronil. In: Proc 5th Int Symp Ectoparas Pets, Fort Collins, Colorado, April 1999, No. 26

Dryden MW, Perez HR, Ulitchny DM (1997) Efficacy of imidacloprid against *Ctenocephalides felis* in dogs and cats under field conditions. In: Proc Bayer Int Flea Control Symp, Satellite Symp BSAVA/WSAVA/FECAVA Conf, Birmingham, April 1997, pp 5-10

Dryden M, Perez HR, Ulitchny DM (1999b) Control of fleas on pets and in homes by use of imidacloprid or lufenuron and a pyrethrin spray. J Am Vet Med Assoc 215:36-39

Dryden MW, Blagburn BL, Bledsoe DL, Denholm I, Hansen O, Hinkle NC, Hopkins T, Jacobs DE, Mehlhorn H, Mencke N, Rust MK, Vaughn MB (2000) Susceptibility of fleas to control agents: development of a monitoring system. In: Proc 45th Ann Meet Am Ass Vet Parasitol, Salt Lahe City, Utah, July 2000, No. 18, p 36

Eckert J, Rommel M, Kutzer E (2000) Erreger von Parasitosen: Systematik, Taxonomie und allgemeine Merkmale. In: Rommel M, Eckert J, Kutzer E, Körting W, Schnieder T (eds) Veterinärmedizinische Parasitologie, 5th edn. Parey Verlag, Berlin, pp 2-39

Edwards FB (1968) Fleas. Vet Rec 85:665

Eidson M, Tierney LA, Rollag OR, Becker T, Brown T, Hull HF (1988): Feline plague in New Mexico: Risk factors and transmission to humans. Am J Public Health 78:1333-1335

Elbel RE (1951) Comparative studies on the larva of certain species of fleas (Siphonaptera). J Parasitol 2:119-128

Elbert A, Nauen R (1996) Bioassays for imidacloprid for a resistance monitoring against the whitefly *Bemisia tabaci*. In: Proc Brighton Crop Protec Conf - Pests and Diseases, pp 731-738

Elbert A, Nauen R (2000) Resistance of *Bemisia* spp. (Homoptera: Aleyrodidae) to insecticides in Southern Spain with special reference to neonicotinoids. Pestic Sci 56:60-64

Elbert A, Becker B, Hartwig J, Erdelen C (1991) Imidacloprid - ein neues systemisches Insektizid (Imidacloprid - a new systemic insecticide). Pflanzenschutz-Nachr. Bayer 44:113-136

Elbert A, Nauen R, Leicht W (1998) Imidacloprid, a novel chloronicotinyl insecticide: Biological activity and agricultural importance. In: Ishaaya I, Degheele D (eds) Insecticides with novel modes of action: mechanism and application. Springer-Verlag, Heidelberg, pp 50-74

Elbert A, Overbeck H, Igawa K, Tsuboi S (1990) Imidacloprid, a novel systemic nitromethylene analogue insecticide for crop protection. In: Proc Brighton Crop Protect Conf - Pests and Diseases, pp 21-28

Elbert A, Nauen R, Cahill M, Devonshire AL, Scarr AW, Sone S, Steffens R (1996) Resistenzmanagement für Chloronikotinylinsektizide am Beispiel von Imidacloprid (Resistance management with chloronicotinyl insecticides using imidacloprid as an example). Pflanzenschutz-Nachr. Bayer 49:5-54

Eldefrawi ME, Eldefrawi AT (1983) Neurotransmitter receptors as targets for insecticides. J Environ Sci Health B18:65-88

El-Gazzar LM, Milio J, Koehler PG, Patterson RS (1986) Insecticide resistance in the cat flea (Siphonaptera: Pulicidae). J Econ Entomol 79:132-134

El-Gazzar LM, Patterson RS, Koehler PG (1988) Comparisons of cat flea (Siphonaptera: Pulicidae) adult and larval insecticide susceptibility. Fla Entomol 71:359-363

Everett R, Cunnigham J, Arther R, Bledsoe DL, Mencke N (2000a) Evaluation of the speed of kill on cat fleas (*C. felis*) on experimentally infested dogs: a comparative study with Advantage® (Imidacloprid) and Revolution™ (Selamectin) In: Proc W Small An Vet Ass / Fed Eu Comp An Vet Ass (WSAVA/FECAVA) Conf, Amsterdam, April 2000, p 503

Everett R, Cunningham J, Arther R, Bledsoe DL, Mencke N (2000b) Comparative evaluation of the speed of flea kill of imidacloprid and selamectin in dogs. Vet Therap 1:229-234

Ewald-Hamm D, Krebber R (1998) Bayer internal report, 5 Feb., 1998, AH-D - ID 18267

Ewald-Hamm D, Krieger KJ, Schein E, Dorn H (1997) Efficacy of Advantage® (imidacloprid) against fleas in naturally infested dogs and cats. Results of two European field studies. In: Proc 4th Int Symp Ectoparas Pets, Riverside, California, April 1997, p 71

Faasch WJ (1935) Darmkanal und Blutverdauung bei Aphaniptera, vor allem *Hystrichopsylla talpae* - Maulwurfsfloh, Maus- und Vogelfloh zum Vergleich. Zeitschr Morph Ök Tiere 29:559-584

Fadok VA (1984) Challenge your clients to gain control of fleas in the environment. Vet Med 79:1039-1044

Fagbemi BO (1982) Effect of *Ctenocephalides felis* strongylus infestation on the performance of West African dwarf sheep and goats. Vet Quart 4:92-95

Farhang-Azad A, Traub R, Sofi M, Wisseman CL (1984) Experimental murine typhus infection in the cat flea, *Ctenocephalides felis* (Siphonaptera: Pulicidae). J Med Entomol 21:675-680

Farnell DR, Faulkner DR (1978) Prepatent period of *Dipetalonema reconditum* in experimentally-infected dogs. J Parasitol 64:565-567

Fehrer SL, Halliwell REW (1987) Effectiveness of Avon's Skin so soft as a flea repellent on dogs. J Am Hosp Assoc 23:217-220

Feingold BF, Benjamini E (1961) Allergy to flea bites: clinical and experimental observations. Ann Allerg 19:1275-1289

Feingold BF, Benjamini E, Michaeli D (1968) The allergic responses to insect bites. Ann Rev Entomol 13:137-158

Ferrari JA (1996) Insecticide resistance. In: Beaty BJ, Marquardt WC (eds) The biology of disease vectors. University Press of Colorado, Niwot, pp 632 ff

Fichtel M (1998) Untersuchungen zur Wirkung des Adultizides Imidacloprid auf den Katzenfloh *Ctenocephalides felis* (Bouché) an Hund und Katze (Investigations of the effect of the adulticide imidacloprid against the cat flea *Ctenocephalides felis* (Bouché) on cats and dogs). Dissertation, Freie Universität Berlin, Fachber Vet Med, Berlin

Fisher MA (1996) Bayer internal report, 27 Apr., 1996, AH-D - ID 16210

Fisher MA, Jacobs DE, Hutchinson MJ, Dick IG (1995) Evaluation of flea control programmes for cats using fenthion and lufenuron. Vet Rec 138:79-81

Fisher MA, Jacobs DE, Hutchinson MJ, McCall JW (2001) Efficacy of imidacloprid on ferrets experimentally infested with the cat flea, *Ctenocephalides felis*. Suppl Comp Cont Educ Vet Pract 23 (4A):8-10

Foil CS (1986) Parasitic dermatoses of the cat. Solvay Vet Dermatol Rep 5 (1):2-3

Forrester DJ, Conti JA, Belden RC (1985) Parasites of the Florida panther (*Felis concolor coryi*). Proc Helminthol Soc Wash 52:95-97

Fox I (1952) Notes on the cat flea in Puerto Rico. Am J Trop Med Hyg 2:337-342

Fox JG (ed) (1998) Biology and diseases of the ferret, 5th edn. Williams and Wilkins, Baltimore

Fox MT (1996) Bayer internal report, 6 Sept., 1996, AH-D - ID 16692

Fox I, Bayona IG (1968) *Alphitobius laevigatus*, a predator on flea larvae. J Econ Entomol 61:877

Fox I, Garcia-Moll I (1961) Ants attacking fleas in Puerto Rico. J Econ Entomol 54:1065-1066

Fox I, Fox RI, Bayona IG (1966) Fleas fed on lizards in the laboratory in Puerto Rico. J Med Entomol 2:395-396

Fox I, Rivera A, Bayona IG (1968) Toxicity of six insecticides to the cat flea. J Econ Entomol 61:869-870

Franc M, Cadiergues MC (1995) Utilisation du lufenuron dans le control des infestations du chien par *Ctenocephalides felis*. Rev Med Vet 146:481-484

Franc M, Cadiergues MC (1998) Antifeeding effect of several insecticidal formulations against *Ctenocephalides felis* on cats. Parasite 5:83-86

Fukase T, Stanneck D, Mencke N (2000) Efficacy and safety of an imidacloprid spot-on formulation for treating flea infestations in domestic rabbits. In: Proc W Small An Vet Ass / Fed Eu Comp An Vet Ass (WSAVA/FECAVA) Conf, Amsterdam, April 2000, p 522

Gaafar SM (1966) Pathogenesis of ectoparasites. In: Soulsby EJ (ed) Biology of parasites. Academic Press, New York, pp 229-236

Galzi JL, Changeux JP (1994) Neurotransmitter-gated ion channels as unconventional allosteric proteins. Curr Opin Struct Biol 4:554-565

Galzi JL, Revah F, Bessis A, Changeux JP (1991) Functional architecture of the nicotinic acetylcholine receptor: from electric organ to brain. Ann Rev Pharmacol Toxicol 31:37-72

Geary JM (1959) The fleas of New York. Cornell Experiment Station Memoir 355, Ithaca

Geary MR (1977) Ectoparasite survey. Brit Vet Dermat Study Group Newsletter 2:2-3

Genchi C (1992) Arthropoda as zoonoses and their implications. Vet Parasitol 44:21-33

Genchi C, Traldi G, Biancardi P (2000) Efficacy of imidacloprid on dogs and cats with natural infestations of fleas, with special emphasis on flea hypersensitivity. Vet Therap 2:71-80

Georghiou GP (1972) The evolution of resistance to pesticides. Ann Rev Ecol Sys 3:133-168

Georghiou GP, Taylor CE (1976) Pesticide resistance as an evolutionary phenomenon. In: Proc 15th Int Cong Entomol, Washington, DC. Entomol Soc Am, 1976, pp 759-785

Georghiou GP, Taylor CE (eds) (1986) Factors influencing the evolution of resistance. In: Pesticide resistance: strategies and tactics for management. National Research Council, National Academy of Sciences Press, Washington, DC., pp 157-169

Georgi JR (ed) (1990) Parasitology for veterinarians, 5th edn. WB Saunders, Philadelphia, pp 200-203

Gordon PL, McEven FL (1984) Insecticide-stimulated reproduction of *Myzus persicae*, the green peach aphid (Homoptera: Aphididae). Can Entomol 116:783-784

Gortel K (1997) Advances in topical and systemic therapy for flea control in dogs. Canine Pract 22:16-21

Gothe R (1985) Pathogenese bei Befall mit Arthropoden. Berl Münch Tierärztl Wochenschr 98:274-279

Grant D (1996) Flea biology and control. Vet Pract 28:7-8

Griffin DL, Canfield PJ, Collins GH (1983) *Ctenocephalides felis felis* infestation of koalas. Aust Vet J 60:275

Griffin L, Krieger K, Liège P (1997) Imidacloprid: A new compound for control of fleas and flea-initiated dermatitis. Suppl Comp Cont Educ Pract Vet 19 (5):17-20

Gross A, Ballivet M, Rungger D, Bertrand D (1991) Neuronal nicotinic acetylcholine receptors expressed in *Xenopus* oocytes: Role of the α subunit in agonist sensitivity and desensitization. Pflügers Arch 419:545-551

Gross TL, Halliwell REW (1985) Lesions of experimental flea bite hypersensitivity in the dog. Vet Pathol 22:78-81

Gundelfinger ED (1992) How complex is the nicotinic receptor system of insects? Trends Neurosci 15:206-211

Guzman RF (1982) *Cheyletiella blakei* (Acari: Cheyletiellidae) hyperparasitic on the cat flea *Ctenocephalides felis felis* (Siphonaptera: Pulicidae) in New Zealand. NZ Entomol 7:322-323

Guzman RF (1984) A survey of cats and dogs for fleas: with particular reference to their role as intermediate hosts of *Dipylidium caninum*. NZ Vet J 32:71-73

Gyr P (1995) Bayer internal report, 13 July, 1995, AH-D - ID 15701

Gyr P, Hopkins T (1995) Bayer internal report, 15 Nov., 1995, AH-D - ID 15962

Gyr P, Kerwick C, Woodley I (1995) Bayer internal report, 22 May, 1995, AH-D - ID 15698

Haarløv N, Kristensen S (1977) Beiträge zur Dermatologie von Hund und Katze. 3. Flöhe von Hunden und Katzen in Dänemark.. Tierärztl Prax 5:507-511

Haas GE (1966) Cat flea - mongoose relationships in Hawaii. J Med Entomol 2:321-326

Haeselbarth E (1966) A note on the subspecies of *Ctenocephalides felis* in Africa south of the Sahara (Siphonaptera: Pulicidae). Zool Anz 176:357-365

Halliwell REW (1979) Flea bite dermatitis. Comp Cont Ed. Pract Vet 1:367-371

Halliwell REW (1981) Hyposensitization in the treatment of flea-bite hypersensitivity: Results of a double blind study. J Am Anim Hosp Assoc 17:249-253

Halliwell REW (1982) Ineffectiveness of thiamine (vitamin B1) as a flea-repellent in dogs. J Am Anim Hosp Assoc 18:423-426

Halliwell REW (1983) Flea allergy dermatitis. In: Kirk RW (ed) Current veterinary therapy VIII. Small Anim Pract. WB Saunders Co., Philadelphia, pp 496-499

Halliwell REW (1984) Managing flea-allergy dermatitis - 3. Factors in the development of flea bite allergy. Vet Med Small Anim Clin 79:1273-1278

Halliwell REW (1985a) Flea allergy: Pathogenesis, therapy, and flea control. In: Proc Am An Hosp Assoc 52nd Meeting, pp 145-149

Halliwell REW (1985b) Personal communication, cited in Kwochka 1987

Halliwell REW (1985c) Personal communication, cited in Moriello 1991

Halliwell REW (1986) Managing canine flea allergy. Solvay Vet Dermatol Rep 5 (1):7-8

Halliwell REW (1990) Clinical and immunological aspects of allergic skin diseases in domestic animals. In: Von Tscharner C, Halliwell REW (eds) Advances in veterinary dermatology, Vol I. Baillière Tindall, London, pp 91-116

Halliwell REW, Longino SJ (1985) IgE and IgG antibodies to flea antigen in differing dog populations. Vet Immunol Immunopath 8:215-223

Halliwell REW, Schemmer KR (1987) The role of basophils in the immunopathogenesis of hypersensitivity to fleas (*Ctenocephalides felis*) in dogs. Vet Immunol Immunopathol 15: 203-213

Halliwell REW, Schwartzman RM (1971) Atopic disease in the dog. Vet Rec 89:209-214

Halliwell RE, Preston JF, Nesbitt JG (1987) Aspects of the immunopathogenesis of flea allergy dermatitis in dogs. Vet Immunol Immunpath 17:483-494

Hanssen I, Mencke N, Asskildt H, Ewald-Hamm D, Dorn H (1999) Field study on the insecticidal efficacy of Advantage against natural infestations of dogs with lice. Parasitol Res 85:347-348

Harman DA, Halliwell RE, Greiner EC (1987) Flea species from dogs and cats in North-Central Florida. Vet Parasitol 23:135-140

Harvey JW, French TW, Meyer DJ (1982) Chronic iron deficiency anemia in dogs. J Am Anim Hosp Assoc 18:946-960

Harwood RF, James MT (eds) (1979) Fleas. In: Entomology in human and animal health, 7th edn. Macmillan, New York, pp 319-341

Heckenberg K, Costa SD, Gregory LM, Michael BF, Endris RG, Shoop WL (1994) Comparison of thumb-counting and comb-counting methods to determine *Ctenocephalides felis* infestation levels on dogs. Vet Parasitol 53:153-157

Heeschen K (1995) Neue Wirkstoffe und Methoden zur Flohbekämpfung bei Hunden (New compounds and methods for control of fleas on dogs). Dissertation, Tierärztliche Hochschule Hannover, Hannover

Heeschen K, Liebisch A (1994) Bayer internal report, 25 April, 1994, AH-D - ID 16010

Hendersen G, Manweiler SA, Lawrence WJ, Templen RJ, Foil LD (1995) The effects of *Steinernema carpocapsae* (Weiser) application to different life stages on adult emergence of the cat flea *Ctenocephalides felis* (Bouché). Vet Dermatol 6:159-163

Hermsen B, Stetzer E, Thees R, Heiermann R, Schrattenholz A, Ebbinghaus U, Kretschmer A, Methfessel C, Reinhardt S, Maelicke A (1998) Neuronal nicotinic receptors in the locust *Locusta migratoria*. J Biol Chem 273:18394-18404

Hewitt M, Walton GS, Waterhouse M (1971) Pet animal infestations and human skin disease. Br J Dermat 86:215-225

Hickey GJ, Chang CH, Marsilio F, Trimboli W, Rickes EL (1993) Effects of prednisone on dermal responses in flea-allergen hypersensitized dogs. Vet Dermatol 4:71-77

Hinaidy HK (1991) Beitrag zur Biologie des *Dipylidium caninum*. 2. Mitteilung (The biology of *Dipylidium caninum*. Part 2). J Vet Med Series B 38:329-336

Hinaidy HK, Bacowsky H, Hinterdorfer F (1987) Einschleppung der Hunde-Filarien *Dirofilaria immitis* und *Dipetalonema reconditum* nach Österreich. J Vet Med B 34:326-332

Hink WF, Drought DC, Barnett S (1991) Effect of an experimental systemic compound, CGA-184699:on life stages of the cat flea (Siphonaptera: Pulicidae). J Med Entomol 28:424-427

Hink WF, Zakson M, Barnett S (1994) Evaluation of a single oral dose of lufenuron to control flea infestation in dogs. Am J Vet Res 55:822-824

Hinkle NC (1992) Biological factors and larval management strategies affecting cat flea (*C. felis felis*) populations. Ph.D. Dissertation, University of Florida, Gainesville

Hinkle NC, Kohler PG, Patterson RS (1995a) Residual effectiveness of insect growth regulators applied to carpet for control of cat flea (Siphonaptera: Pulicidae) larvae. J Econ Entomol 88:903-906

Hinkle NC, Rust MK, Reierson DA (1997) Biorational approaches to flea (Siphonaptera: Pulicidae) suppression: present and future. J Agric Entomol 14:309-321

Hinkle NC, Sheppard DC, Bondari K, Butler JF (1989) Effect of temperature on toxicity of three pyrethroids to horn flies. Med Vet Entomol 3:435-439

Hinkle NC, Wadleigh RW, Koehler PG, Patterson RS (1995b) Mechanisms of insecticide resistance in a strain of cat fleas (Siphonaptera: Pulicidae). J Entomol Sci 30:43-48

Hopkins D (1980) Ectoparasites of the Virginia opossum (*Didelphis virginiana*) in an urban environment. Northwest Sci 54:199-201

Hopkins GHE, Rothschild M (1953) An illustrated catalogue of the Rothschild collection of fleas (Siphonaptera) in the British Museum (Natural History). Vol. I-IV. University Press, Cambridge

Hopkins TJ (1995) Bayer internal report, 21 Feb., 1995, AH-D - ID 17100

Hopkins TJ (1997a) Flea control in Australia. Efficacy of a new compound, imidacloprid, against *Ctenocephalides* spp. in dogs and cats. In: 16th Int Conf W Ass Adv Vet Parasitol, Sun City, August 1997, No. 146, p 40

Hopkins TJ (1997b) Imidacloprid: *Ctenocephalides felis* control on dogs under field conditions in Australia. Suppl Comp Cont Educ Pract Vet 19 (5):25-26

Hopkins TJ (1997c) Imidacloprid: *Ctenocephalides felis* control on dogs under field conditions in Australia. In: Proc Bayer Int Flea Control Symp, Satellite Symp BSAVA/WSAVA/FECAVA Conf, April 1997, Birmingham, pp 13-16

Hopkins TJ (1998) Imidacloprid and resolution of signs of flea allergy dermatitis in dogs. Canine Pract 23:18-23

Hopkins TJ, Kerwick C, Gyr P, Woodley I (1996a) Efficacy of imidacloprid to remove and prevent *Ctenocephalides felis* infestations on dogs and cats. Aust Vet Practit 26:150-153

Hopkins TJ, Kerwick C, Gyr P, Woodley I (1997a) Efficacy of imidacloprid to remove and prevent *Ctenocephalides felis* infestations on dogs and cats. Suppl Compend Cont Educ Pract Vet 19 (5):11-16

Hopkins TJ, Woodley I, Gyr P (1996b) Imidacloprid topical formulation: larvicidal effect against *Ctenocephalides felis* in the surroundings of treated dogs. Aust Vet Pract 26:210

Hopkins TJ, Woodley I, Gyr P (1997b) Imidacloprid topical formulation: larvicidal effect against *Ctenocephalides felis* in the surroundings of treated dogs. Suppl Comp Cont Educ Pract Vet 19 (5):4-10

Horak IG (1982) Parasites of domestic and wild animals in South Africa. XIV. The seasonal prevalence of *Rhipicephalus sanguineus* and *Ctenocephalides* spp. on kennelled dogs in Pretoria North. Onderstepoort J Vet Res 49:63-68

Horowitz AR, Ishaaya I (1997) Managing resistance to novel insecticides: the Israeli experience. Phytoparasitica 25:346-347

Hoskins WM, Gordon HAT (1956) Arthropod resistance to chemicals. Ann Rev Entomol 1:89-123

Huang Y, Williamson MS, Devonshire AL, Windass JD, Lansdell SJ, and Millar NS (1999) Molecular characterization and imidacloprid selectivity of nicotinic acetylcholine receptor subunits from peach-potato aphid *Myzus persicae*. J Neurochem 73:380-389

Hudson BW, Prince FM (1958) A method for large-scalerearing of the cat flea, *Ctenocephalides felis felis* (Bouché). Bull WHO 19:1126-1129

Hudson BW, Feingold BF, Kartman L (1960) Allergy to flea bites. I. Experimental induction of flea-bite sensitivity in guinea pigs. Exp Parasitol 9:18-24

Hunter KW, Campbell AR, Sayles PC (1979) Human infestation by cat fleas, *Ctenocephalides felis* (Siphonaptera: Pulicidae), from suburban raccoons. J Med Entomol 16:547

Hutchinson MJ, Jacobs DE, Bell GD, Mencke N (2001a) Evaluation of imidacloprid for the treatment and prevention of cat flea (*Ctenocephalides felis felis*) infestation on rabbits. Vet Rec (in press)

Hutchinson MJ, Jacobs DE, Mencke N (2001b) Establishment of the cat flea (*Ctenocephalides felis felis*) on the ferret (*Mustela putorius furo*) and its control with imidacloprid. Med Vet Entomol (in press)

Ioff I, Pokrovskaya M (1929) Experiments with fleas of human dwellings as carriers of plague infection. Izv Gos Mikrobiol Inst Rostov-na-Donu 9:126-136

Irons JV, Beck O, Murphy JN (1946) Fleas carrying typhus Rickettsiae found on nonmurine hosts. J Bacteriol 51:609-610

Jacobs DE (1995) Bayer internal report, 6 Dec., 1995, AH-D - ID 15749

Jacobs DE (2000) Adulticidal and larvicidal effects of imidacloprid: Two-stage control. Suppl Comp Cont Educ Pract Vet 22 (4A):15-17

Jacobs DE (2001) Efficacy of imidacloprid (Advantage®) on rabbits naturally or experimentally infested with the cat flea (*Ctenocephalides felis*). In: Proc North Am Vet Conf, Orlando, Florida, January 2001, Vol. 15, pp 485-486

Jacobs DE, Hutchinson MJ (1998) Bayer internal report, 8 May, 1998, AH-D - ID 18984

Jacobs DE, Hutchinson MJ , Ewald-Hamm D (2000) Inhibition of immature *Ctenocephalides felis felis* (Siphonaptera: Pulicidae) development in the immediate environment of cats treated with imidacloprid. J Med Entomol 37:228-230

Jacobs DE, Hutchinson MJ, Fukase T, Hansen O (2001) Efficacy of imidacloprid on rabbits naturally or experimentally infested with the cat flea, *Ctenocephalides felis*. Suppl Comp Cont Educ Pract Vet 23 (4A):11-14

Jacobs DE, Hutchinson MJ, Krieger KJ (1997a) Duration of activity of imidacloprid, a novel adulticide for flea control, against *Ctenocephalides felis* on cats. Vet Rec 140, 259-260

Jacobs DE, Hutchinson MJ, Fox MT, Krieger KJ (1997b) Comparison of flea control strategies using imidacloprid or lufenuron on cats in a controlled simulated home environment. Am J Vet Res 58:1260-1262

Jenkins DW (1964) Pathogens, parasites, and predators of medically important arthropods. Annoted list and bibliography. Bull WHO Suppl 30:86

Joseph SA (1974) *Ctenocephalides felis orientis* Jordan, 1925 as an intermediate host of the tapeworm *Dipylidium caninum* (Linn 1758) Railliet 1892. Cheiron 3:70-75

Joseph SA (1976) Observations on the feeding habits of *C. felis orientis* (Jordan, 1925) on human hosts. Cheiron 5:73-77

Joseph SA (1981) Studies on the bionomics of *Ctenocephalides felis orientis* (Jordan 1925). Cheiron 10:275-280

Joseph SA (1985) Flea infestation in horses and in syces. Centaur 2:59-60

Joseph SA, Karunamoorthy G, Lalitha CM (1984) Cat flea, *Ctenocephalides felis felis* (Bouché), infestation in a poultry farm of Tamil Nadu. Indian J Poultry Sci 19:192-193

Joyeux C (1916) Sur le cycle evolutif de quelques Cestodes. Bull Soc Pathol Exot 9:578-583

Kagabu S (1996) Studies on the synthesis and insecticidal activity of neonicotinoid compounds. Pestic Sci 46:231-239

Kagabu S (1997) Molecular structures and properties of imidacloprid and analogous compounds. Phytoparasitica 25:347

Kagabu S, Medej S (1995) Stability comparison of imidacloprid and related compounds under simulated sunlight, hydrolysis conditions, and to oxygen. Biosci Biotech Biochem 59:980-985

Kalkofen UP, Greenberg J (1974) Public health implications of *Pulex irritans* infestations of dogs. J Am Vet Med Assoc 165:903-905

Kalvelage H, Münster M (1991) Die *Ctenocephalides-canis-* und *Ctenocephalides-felis-*Infestation von Hund und Katze. Tierärztl Praxis 19:200-206

Karandikar KR, Munshi DM (1950) Life history and bionomics of the cat flea, *Ctenocephalides felis felis* (Bouché). J Bombay Nat His Soc 49:169-177

Kaufmann AF, Mann JM, Gardiner TM, Heaton F, Poland JD, Barnes AM, Maupin GO (1981) Public health implications of plague in domestic cats. J Am Vet Med Assoc 179:875-878

Keaton R, Nash BJ, Murphy JN, Irons JV (1953) Complement fixation tests for murine typhus on small mammals. Publ Health Rep. 68:28-30

Keep KM (1983) Flea allergy dermatitis. New South Wales Vet Proc 19:24-27

Kern WH (1991) Cat flea larva - the unknown life stage. Pest Manage 10:20-22

Kern WH Jr, Koehler PG, Patterson RS (1992) Diel patterns of cat flea (Siphonaptera: Pulicidae) egg and fecal deposition. J Med Entomol 29:203-206

Kern WH Jr, Richman DL, Koehler PG, Brenner RJ (1999) Outdoor survival and development of immature cat fleas (Siphonaptera: Pulicidae) in Florida. J Med Entomol 36:207-211

Kerr RW (1977) Resistance to control chemicals in Australian arthropod pests. J Aust Entomol Soc 16:312-334

Kieffer M, Kristensen S (1979) Flea hypersensitivity in dogs and cats. Int J Dermatol 18:707-712
Kieffer M, Kristensen S, Hallas TE (1979) Prurigo and pets: The benefit from vets. Br Med J 1 (6177):1539-1540
Kissileff A (1938) The dog flea as a causative agent in summer eczema. J Am Vet Med Assoc 93:21-27
Kissileff A (1969) Flea-bite dermatitis. Vet Med Small Anim Clin 64:580-583
Kissileff A (1987) Letter to the editor: Flea control. J Am Vet Med Assoc 190:1372
Knaust HJ, Poehling HM (1992) Untersuchungen zur Wirkung von Imidacloprid auf Getreideblattläuse und deren Effizienz zur Übertragung des BYD-Virus (Studies of the action of imidacloprid on grain aphids and their efficiency to transmit BYD-Virus). Pflanzenschutz-Nachr Bayer 45:381-408
Knipling EF, Klassen W (1984) Influence of insecticide use patterns on the development of resistance to insecticides - A theoretical study. Southwestern Entomologist 9:351-368
Kobayashi Y, Ono Y, Okano T, Buéi K (1994) Insecticide susceptibility of *Ctenocephalides felis* (Siphonaptera: Pulicidae) to malathion and permethrin in Tanzania. Med Vet Entomol 2:325-329
Koehler PG, Leppla NC, Patterson RS (1989) Circadian rhythm of cat flea (Siphonaptera: Pulicidae) locomotion unaffected by ultrasound. J Econ Entomol 82:516-518
Koehler PG, Milio J, Patterson RS (1986) Residual efficacy of insecticides applied to carpet for control of cat fleas (Siphonaptera: Pulicidae). J Econ Entomol 79:1036-1038
Kollmeyer WD, Flattum RF, Foster JP, Powell JE, Schroeder ME, Soloway SB (1999) Discovery of the nitromethylene heterocycle insecticides. In: Yamamoto I, Casida JE (eds) Nicotinoid insecticieds and nicotinic acetylcholine receptor. Springer-Verlag, Tokyo, pp 71-91
Krieger KJ (1996) Bayer internal report, 12 Jan., 1996, AH-D - ID 15790
Kristensen S (1978) Beiträge zur Dermatologie von Hund und Katze. 4. Flohallergie. Tierärztl Prax 6:351-360
Kristensen S, Kieffer M (1978) A study of skin diseases in dogs and cats. V. The intradermal test in the diagnosis of flea allergy in dogs and cats. Nord Vet Med 30:414-423
Kristensen S, Haarløv N, Mourier H (1978) A study of skin diseases in dogs and cats. IV. Patterns of flea infestation in dogs and cats in Denmark. Nord Vet Med 30:401-413
Kunkle G (1997) An updated review of ectoparasiticide treatments in dogs and cats. J Vet Pharmacol Therap 20 (Suppl 1):110-120
Kunkle GA, Milcarsky J (1985) Double-blind flea hyposensitization trial in cats. J Am Vet Med Assoc 186:677-680
Kunz SE, Kemp DH (1994) Insecticides and acaricides: resistance and environmental impact. Rev Sci Tech Off Ont Epiz 13:1249-1286
Kwochka KW (1987) Fleas and related disease. Vet Clin N. Am Small Anim Pract 17:1235-1262
Kwochka KW, Bevier DE (1987) Flea dermatitis. In: Nesbitt GH (ed) Contemporary issues in small animal practice. Dermatology. Vol. 8. Churchill Livingstone, New York, pp 21-55
Lavoipierre MMJ, Hamachi M (1961) An apparatus for the observations on the feeding mechanism of the flea. Nature 192:998-999
Le Novere N, Changeux JP (1995) Molecular evolution of the nicotinic acetylcholine receptor: and example of multigene family in excitable cells. J Mol. Evol 40:155-172
Leicht W (1993) Imidacloprid - a chloronicotinyl insecticide. Pesticide Outlook 4:17-21
Leicht W (1996) Imidacloprid - ein Chloronicotinyl-Insektizid: Biologische Aktivität und landwirtschaftliche Bedeutung (Imidacloprid - a chloronicotinyl insecticide: biological activity and agricultural significance). Pflanzenschutz-Nachr Bayer 49:71-86
Lemke LA, Koehler PG, Patterson RS (1989) Susceptibility of the cat flea (Siphonaptera: Pulicidae) to pyrethroids. J Econ Entomol 82:839-841
Lewis RE (1972) Notes on the geographic distribution and host preferences in the order Siphonaptera. Part 1. Pulicidae. J Med Entomol 9:511-520
Lewis RE, Lewis JH, Maser C (1988) The fleas of the Pacific Northwest. Oregon State University Press, Corvallis
Liberg O, Sandell M (1988) Spatial organization and reproductive tactics in the domestic cat and other felids In: Turner CE (ed) The domestic cat: the biology of its behaviour. Cambridge Univ Press, pp 83-98

Liebisch A, Heeschen K (1995) Bayer internal report, 26 June, 1995, AH-D - ID 16049

Liebisch A, Heeschen K (1997) Controlled laboratory study on the efficacy of imidacloprid spot-on for control of the cat flea in dogs. In: Proc Bayer Int Flea Control Symp, Satellite Symp BSAVA/WSAVA/FECAVA Conf, Birmingham, April 1997, pp 25-28

Liebisch A, Krebber R (1997) Bayer internal report, 24 March, 1997, AH-D - ID 17719

Liebisch A, Reimann U (2000) The efficacy of imidacloprid against flea infestation on dogs compared with three other topical preparations. Canine Pract 25:8-11

Liebisch A, Brandes R, Hoppenstedt K (1985) Zum Befall von Hunden und Katzen mit Zecken und Flöhen in Deutschland (Tick and flea infections of dogs and cats in the German Federal Republic). Prakt Tierarzt 66:817-824

Linardi PM, Nagem RL (1972) Observações sobre o ciclo evolutivo de Ctenocephalides felis (Bouché, 1835) (Siphonaptera: Pulicidae) e sua sobrevida fora do hospedeiro. Bol Mus Hist Nat Zool Universidade Federal de Minas gerais Belo Horizonte 13:1-22

Linardi PM, Demaria M, Botelho JR (1997) Effects of larval nutrition on the postembryonic development of Ctenocephalides felis felis (Siphonaptera: Pulicidae). J Med Entomol 34:494-497

Lind RJ, Clough MS, Reynolds SE, Earley FGP (1998) [3H]Imidacloprid labels high- and low-affinity nicotinic acetylcholine receptor-like binding sites in the aphid Myzus persicae (Hemiptera: Aphididae). Pestic Biochem Physiol 62:3-14

Liu MY, Casida JE (1993) High affinity binding of [3H]imidacloprid in the insect acetylcholine receptor. Pestic Biochem Physiol 46:40-46

Liu MY, Lanford J, Casida JE (1993) Relevance of [3H]imidacloprid binding site in house fly head acetylcholine receptor to insecticidal activity of 2-nitromethylene- and 2-nitroimino-imidazolidines. Pestic Biochem Physiol 46:200-206

Logas DB (1995) The cat, the flea, and pesticides. Vet Clin N Am Small Anim Pract 25:801-811

Lombarddero OJ, Santa-Cruz AM (1986) Parasites of stray dogs in the city of Corrientes (Argentina). Changes over a 25 year-period. Vet Argent 3:888-892

Londershausen M (1996) Approaches to new parasiticides. Pestic Sci 48:269-292

Lorenz MD (1980) The management of flea allergy dermatitis. In: Kirk RW (ed) Current Veterinary Therapy VII. Small Anim Pract. WB Saunders Co., Philadelphia, pp 446-450

Lowery DT, Sears MK (1986) Stimulation of reproduction of the green peach aphid (Homoptera: Aphididae) by azinphosmethyl applied to potatoes. J Econ Entomol 79:1530-1533

Luetje W, Patrick J (1991) Both alpha- and beta-subunits contribute to the agonist sensitivity of neuronal nicotinic acetylcholine receptors. J Neurosci 11:837-845

Luetje W, Wada K, Rogers S, Abramson SN, Tsuji K, Heinemann S, Patrick J (1990) Neurotoxins distinguish between different neuronal nicotinic acetylcholine receptor subunit combinations. J Neurochem 55:632-640

Lukas RJ, Bencherif M (1992) Heterogeneity and regulation of nicotinic acetylcholine receptors. Int Rev Neurobiol 34:25-131

Lummis SCR, and Sattelle DB (1985) Binding of N-[propionyl-3H]propionylated alpha-bungarotoxin and L-[benzilic-4,4'-3H]quinudidinyl benzilate to CNS extracts of the cockroach Periplaneta americana. Comp Biochem Physiol C 80:75-83

Lyons H (1915) Notes on the cat flea (Ctenocephalides felis (Bouché)). Psyche 22:124-132

MacDonald JM (1983) Ectoparasites. In: Kirk RW (ed) Current veterinary therapy VIII. Small Anim Pract. WB Saunders Co, Philadelphia, pp 488-495

MacDonald JM (1984) Managing flea-allergy dermatitis - 3. Solving the Southeastern triad. Vet Med Small Anim Clin 79:1278-1280

MacDonald JM (1995) Flea control: An overview of treatment concepts for North America. Vet Dermatol 6:121-130

MacDonald JM, Miller TA (1986) Parasiticide therapy in small animal dermatology. In: Kirk RW (ed) Current veterinary therapy IX. Small Anim Pract. WB Saunders Co, Philadelphia, pp 571-590

Marchiondo AA, Green SE, Plue RE, Wallace DH, Barrick RA, Jeannin P (1999) Comparative speed of kill of Frontline® spray, Frontline® Top-Spot™ and Advantage® against adult cat fleas (Ctenocephalides felis) on dogs. In: Proc 5th Int Symp Ectoparas Pets, Fort Collins, Colorado, April 1999, No. 30

Marsella R (1999) Advances in flea control. Vet Clin N. Am Small Anim Pract 29:1407-1424

Marshall AG (1967) The cat flea, *Ctenocephalides felis felis* (Bouché, 1835) as an intermediate host for cestodes. Parasitology 57:419-430

Marshall AG (1981) The ecology of ectoparasitic insects. Academic Press, London, New York

Mason IS (1998) Bayer internal report, 11 Feb., 1998, AH-D - ID 19409

Matsuda K, Buckingham SD, Freeman JC, Squire MD, Baylis HA, Sattelle DB (1999) Role of the α subunit of nicotinic acetylcholine receptor in the selective action of imidacloprid. Pestic Sci 55:211-213

Maunder JH (1984) Entomological components to urticaria. Practitioner 228:1051-1055

McGehee DS, Role LW (1995) Physiological diversity of nicotinic acetylcholine receptors expressed by vertebrate neurones. Ann Rev Physiol 57:521-546

Mehlhorn H (2000) Mode of action of imidacloprid and comparison with other insecticides (i.e., fipronil and selamectin) during in vivo and in vitro experiments. Suppl Comp Cont Educ Pract Vet 22 (4A):4-8

Mehlhorn H, Mencke N, Hansen O (1999) Effects of imidacloprid on adult and larval stages of the cat lea *Ctenocephalides felis* after in vivo and in vitro application: a light- and electron-microscopy study. Parasitol Res 85:625-637

Mehlhorn H, D'Haese J, Mencke N, Hansen O (2001a) In vivo and in vitro effects of imidacloprid on sheep keds (*Melophagus ovinus*): a light and electron microscopic study. Parasitol Res (in press)

Mehlhorn H, Hansen O, Mencke N (2001b) Comparative study on the effects of three insecticides (fipronil, imidacloprid, selamectin) on the developmental stages of the cat flea (*Ctenocephalides felis* Bouché 1835): a light and electron microscopic analysis of in vivo and in vitro experiments. Parasitol Res (in press)

Ménier K, Beaucournu JC (1998) Taxonomic study of the genus *Ctenocephalides* Stiles & Collins, 1930 (Insecta: Siphonaptera: Pulicidae) by using aedegous characters. J Med Entomol 35:883-890

Merchant SR (1990) Zoonotic diseases with cutaneous manifestations - Part I. Comp Cont Edu Pract Vet 12 (3):371-377

Methfessel C (1992) Die Wirkung von Imidacloprid am nikotinergen Acetylcholin-Rezeptor des Rattenmuskels (Action of imidacloprid on the nicotinergic acetylcholine receptors in rat muscle). Pflanzenschutz-Nachr Bayer 45:369-380

Metzger ME (1995) Photoperiod and temperature effects on the development of *Ctenocephalides felis* (Bouché) and studies on its chemical control in turfgrass. MS Thesis, University of California, Riverside

Metzger ME, Rust MK (1996) Egg production and emergence of adult cat fleas (Siphonaptera: Pulicidae) exposed to different photoperiods. J Med Entomol 33:651-655

Metzger ME, Rust MK (1997) Effect of temperature on cat flea (Siphonaptera: Pulicidae) development and overwintering. J Med Entomol 34:173-178

Michaeli D, Benjamini E, Deburen FP, Larrivee DH, Feingold BF (1965) The role of collagen in the induction of flea bite hypersensitivity. J Immunol 95:162-170

Moffat AS (1993) New chemicals seek to outwit insect pests. Science 261:550-551

Moriello KA (1987) Common ectoparasites of the dog. Part I: Fleas and ticks. Canine Pract 14:6-18

Moriello KA (1991) Parasitic hypersensitivity. Sem Vet Med Surg (Sm Anim) 6:286-289

Moriello KA, McMurdy MA (1989a) Feline flea allergy dermatitis - Practice tips on making a diagnosis. Comp Anim Pract 19 (4&5):23-27

Moriello KA, McMurdy MA (1989b) The prevalence of positive intradermal skin test reactions to flea extract in clinically normal cats. Comp Anim Pract 19 (3):28-30

Morlan HB (1952) Host relationships and seasonal abundance of some southwest Georgia ectoparasites. Am Midl Nat 48:74-93

Moser BA, Koehler PG, Patterson RS (1991) Effect of larval diet on cat flea (Siphonaptera: Pulicidae) developmental times and adult emergence. J Econ Entomol 84:1257-1261

Moyses EW (1995) Measurement of insecticide resistance in the adult flea. In: Proc 3rd Int Symp Ectoparas Pets, College Station, Texas, April 1995, pp 21-34

Moyses EW (1997) A comparative study of two laboratory insecticide bioassays for the cat flea. In: Proc 4th Int Symp Ectoparas Pets, Riverside, California, April 1997, p 41

Moyses EW, Buchy A (1996) Rapid selection of malathion resistance in the cat flea *Ctenocephalides felis* (Bouché) (Siphonaptera: Pulicidae). In: Proc 20th Int Cong Entomol, Firenze, August 1996, p 596

Mulle C, Vidal C, Benoit P, Changeux JP (1991) Existence of different subtypes of nicotinic acetylcholine receptors in the rat habenulo-interpenduncular system. J Neurosci 11:2588-2597

Muller GH, Kirk RW (eds) (1976a) Flea allergy dermatitis (canine). In: Small animal dermatology, 2nd edn. WB Saunders Co, Philadelphia, pp 403-408

Muller GH, Kirk RW (eds) (1976b) Flea allergy dermatitis (feline), miliary dermatitis (feline eczema). In: Small animal dermatology, 2nd edn. WB Saunders Co, Philadelphia, pp 409-412

Muller GH, Kirk RW, Scott DW (eds) (1983) Parasitic hypersensitivity: Flea allergy dermatitis. In: Small animal dermatology, 3rd edn. WB Saunders Co, Philadelphia, pp 432-440

Müller J, Kutschmann K (1985) Flohnachweise (Siphonaptera) auf Hunden im Einzugsbereich der Magdeburger Poliklinik für kleine Haus- und Zootiere (Records of fleas (Siphonaptera) on pets and small zoo animals). Angew Parasitol 26:197-203

Mullins JW (1993) Imidacloprid - A new nitroguanidine insecticide. ACS Symp, USA, 1992, Series No. 524, Pest control with enhanced environmentals safety. Washington, DC, 1993, 203, pp 183-198

Nagata K, Aistrup GL, Song JH, Narahashi T (1996) Subconductance-state currents generated by the nitromethylene heterocycle imidacloprid at the nicotinic acetylcholine receptor in PC12 cells. Neuroreport 7:1025-1028

Nagata K, Iwanaga Y, Shono T, Narahashi T (1997) Modulation of the neuronal acetylcholine receptor channel by imidacloprid and cartap. Pestic Biochem Physiol 59:119-128

Nagata K, Song JH, Shono T, Narahashi T (1998) Modulation of the neuronal nicotinic acetylcholine receptor-channel by the nitromethylene heterocycle imidacloprid. J Pharmacol Exp Therap 285:731-738

Narahashi T (1996) Neuronal ion channels as the target sites of insecticides. Pharmacol Toxicol 78:1-14

Nauen R (1995) Behaviour modifying effects of low systemic concentrations of imidacloprid on *Myzus persicae* with special reference to an antifeeding response. Pestic Sci 44:145-153

Nauen R, Elbert A (1997) Apparent tolerance of a field-collected strain of *Myzus nicotianae* to imidacloprid due to strong antifeeding responses. Pestic Sci 49:252-258

Nauen R, Ebbinghaus-Kintscher U, Elbert A, Jeschke P, Tietjen K (2001) Acetylcholine receptors as sites for developing neonicotinoid insecticides. In: Ishaaya I (ed) Biochemical sites important in insecticide action and resistance. Springer-Verlag, Berlin, Heidelberg, pp 77-105

Nauen R, Ebbinghaus U, Tietjen K (1999) Ligands of the nicotinic acetylcholine receptor as insecticides. Pestic Sci 55:608-610

Nauen R, Hungenberg H, Tollo B, Tietjen K, Elbert A (1998b) Antifeedant effect, biological efficacy and high affinity binding of imidacloprid to acetylcholine receptors in *Myzus persicae* and *Myzus nicotianae*. Pestic Sci 53:133-140

Nauen R, Koob B, Elbert A (1998a) Antifeedant effects of sublethal dosages of imidacloprid on *Bemisia tabaci*. Entomol Experiment Applic 88:287-293

Nauen R, Koob B, Klüver T, Elbert A (1997) Biochemical characterization of insecticide resistant strains of the tobacco whitefly *Bemisia tabaci* (Homoptera: Aleyrodidae). Mittlg Dtsch Gesellsch Allg Angew Entomol 11:217-222

Naumann K (1994) Neue Insektizide (New insecticides). Nachr Chem Tech Lab ('Blaue Blätter') 42:255-262

Nelson GS (1962) *Dipetalonema reconditum* (Grassi, 1889) from the dog with a note on its development in the flea, *Ctenocephalides felis* and the louse, *Heterodoxus spiniger*. J Helminthol 36:297-308

Nelson GS, Heisch RB, Furlong M (1962) Studies in filariasis in East Africa. II. Filarial infections in man, animals and mosquitoes on the Kenya Coast. Trans R Soc Trop Med Hyg 56:202-217

Nesbitt GH (1978) Canine allergic inhalant dermatitis: A review of 230 cases. J Am Vet Med Assoc 172:55-60

Nesbitt GH (1983) Parasitic diseases. In: Canine and feline dermatology: a systematic approach. Lea & Febiger, Philadelphia, pp 65-80

Nesbitt GH, Schmitz JA (1977) Contact dermatitis in the dog: A review of 35 cases. J Am Anim Hosp Assoc 13:155-163

Nesbitt GH, Schmitz JA (1978) Flea bite allergic dermatitis. A review and survey of 330 cases. J Am Vet Med Assoc 173:282-288

Nishimura K, Kanda Y, Okazawa A, Ueno T (1994) Relationship between insecticidal and neurophysiological activities of imidacloprid and related compounds. Pestic Biochem Physiol 50:51-59

Obasaju MF, Otesile EB (1980) *Ctenocephalides felis* infestation of sheep and goats. Trop Anim Health Prod 12:116-118

Okazawa A, Akamatsu M, Ohoka A, Nishiwaki H, Cho WJ, Nakagawa Y, Nishimura K, Ueno T (1998) Prediction of the binding mode of imidacloprid and related compounds to house-fly head acetylcholine receptors using three-dimensional QSAR analysis. Pestic Sci 54:134-144

Olsen A (1982) Årsberetning Annual Report 1982. Danish Infestation Laboratory, Lyngby, Denmark

Olsen A (1985) Ovicidal effect on the cat flea, *Ctenocephalides felis* (Bouché), of treating fur of cats and dogs with methoprene. Int Pest Control 27:10-13, 16

Ortells MO, Lunt GG (1995) Evolutionary history of the ligand-gated ion-channel superfamily of receptors. Trends Neurosci 18:121-127

Osbrink WLA, Rust MK (1984) Fecundity and longevity of the adult cat flea, *Ctenocephalides felis felis* (Siphonaptera: Pulicidae). J Med Entomol 21:727-731

Osbrink WLA, Rust MK (1985a) Seasonal abundance of adult cat fleas, *Ctenocephalides felis* (Siphonaptera: Pulicidae) on domestic cats in southern California. Bull Soc Vector Ecol 10:30-35

Osbrink WLA, Rust MK (1985b) Cat flea (Siphonaptera: Pulicidae): Factors influencing host-finding behaviour in the laboratory. Ann Entomol Soc Am 78:29-34

Osbrink WLA, Rust MK, Reierson DA (1986) Distribution and control of cat fleas in houses in Southern California (Siphonaptera: Pulicidae). J Econ Entomol 79:135-140

Oxenham M (1996) Flea control in ferrets. Vet Rec 138:372

Painter HF, Echerlin RP (1985) The status of the dog flea. Va J Sci 36:114

Papke RL, Boulter J, Patrick J, Heinemann S (1989) Single-channel currents of rat neuronal nicotinic acetylcholine receptors expressed in *Xenopus* oocytes. Neuron 3:589-596

Patton WS (1931) Insects, ticks, mites and venomous animals of medical and veterinary importance. Part II. Public Health Brugg, Great Britain

Paul A, Jones C (1997) Comparative evaluation of imidacloprid and lufenuron for flea control on dogs in a controlled simulated home environment. Suppl Comp Cont Educ Pract Vet 19 (5):35-37

Paul AJ, Jones CJ, Arther RG (1997) Comparative evaluation of Advantage® (imidacloprid) and Program® (lufenuron) for flea control on dogs in a controlled simulated home environment. In: Proc 4th Int Symp Ectoparas Pets, Riverside, California, April 1997, p 75

Pennington NE, Phelps CA (1969) Canine filariasis on Okinawa, Ryuku Islands. J Med Entomol 6:59-67

Persson L (1973) Ektoparasiter hos hund och katt. Svensk Veterinartidn 25:254-260

Petter F (1973) Les animaux domestiques et leurs ancêtres. Bordas Edition, Paris

Pflüger W, Schmuck R (1991) Profil der ökologischen Wirkungen von Imidacloprid (Ecotoxicological profile of imidacloprid). Pflanzenschutz Nachr Bayer 44:145-158

Pickens LG, Carroll JF, Azad AF (1987) Electrophysiological studies of the spectral sensitivities of cat fleas, *Ctenocephalides felis*, and oriental rat fleas, *Xenopsylla cheopsis* to monochromatic light. Entomol Exp Appl 45:193-204

Pinter L (1999) *Leporacarus gibbus* and *Spilopsyllus cuniculi* infestation in a pet rabbit. J Small Anim Pract 40, 220-221

Piotrowski F, Polomska J (1975) Ectoparasites of the dog (*Canis familiaris* L.) in Gdansk. Wiad Parazytol 21:441-451

Pollitzer R (1960) A review of recent literature on plague. Bull WHO 23:397-400

Pollmeier M, Barth D, Jeannin P (1999) A comparison of the flea control efficacy of Frontline® Top Spot™ for cats against the "cottontail" strain and a reference strain. In: Proc 5th Int Symp Ectoparas Pets, Fort Collins, Colorado, April 1999, No. 31

Pospischil R (1995) Influence of temperature and relative humidity on the development of the cat flea (*Ctenocephalides felis*). (In: Proc 16th Tagung Dtsch Gesellsch Parasitol, März 1994, pp 54-69) Zbl Bakt 282:193-194

Prabhaker N, Toscano NC, Castle SJ, Henneberry TJ (1997) Selection for imidacloprid resistance in silverleaf whiteflies from the imperial valley and development of a hydroponic bioassay for resistance monitoring. Pestic Sci 51:419-42

Prichard R (1994) Anthelmintic resistance. Vet Parasitol 54:259-268

Pugh RE (1987) Effects on the development of *Dipylidium caninum* and on the host reaction to this parasite in the adult flea (*Ctenocephalides felis felis*). Parasitol Res 73:171-177

Pugh RE, Moorhouse DE (1985) Factors affecting the development of *Dipylidium caninum* in *Ctenocephalides felis felis* (Bouché 1835). Z Parasitenkd 71:765-775

Pullen SR, Meola RW (1995) Survival and reproduction of the cat flea (Siphonaptera: Pulicidae) fed human blood on an artificial membrane system. J Med Entomol 32:467-470

Rak H (1972) *Cheyletiella parasitivorax* (Megnin) (Acar., Cheyletiellidae) from cat flea (*Ctenocephalides felis felis* (Bouché)) in Iran. Entomol Mon Mag 108:62

Reed WT, Erlam GJ (1978) The housefly metabolism of nitromethylene insecticides. In: Shankland DL, Hollingworth RM, Smyth T Jr (eds) Pesticide and venom neurotoxicity. Plenum, New York, pp 159-169

Reedy L (1977) Ectoparasites. In: Kirk RW (ed) Current veterinary therapy VI. Small Anim Pract. WB Saunders Co, Philadelphia, pp 547-554

Reedy LM (1975) The use of flea antigen in treatment of feline flea-allergy dermatitis. Vet Med Small Anim Clin 70:703-704

Reedy LM (1986) Common parasitic problems in small animal dermatology. J Am Vet Med Assoc 188:362-364

Reedy LM, Miller WH (eds) (1989a) Allergy testing. In: Allergic skin diseases of dogs and cats. WB Saunders Co, Philadelphia, pp 81-109

Reedy LM, Miller WH (eds) (1989b) Fleas and flea bite hypersensitivity. In: Allergic skin diseases of dogs and cats. WB Saunders Co, Philadelphia, pp 171-187

Reese D (1981) Guidelines for flea control. Pest Control May, 19-23

Rehacek J, Fischer RG, Luecke DH (1973) Friend Leukemia Virus (FLV) activity in certain arthropods. II Quantitation infectivity determinations. Neoplasma 20:147-158

Ressl F (1963) Die Siphonapterenfauna des Verwaltungsbezirkes Scheibbs (Niederösterreich). Z Parasitenk 23:470-490

Richman DL, Koehler PG (1997) Effect of temperature and the synergist piperonyl butoxide on imidacloprid toxicity to cat fleas (*Ctenocephalides felis felis*). In: Proc 4th Int Symp Ectoparas Pets, Riverside, California, April 1997, p 77-79

Richman DL, Koehler PG, Brenner RJ (1999) Effect of temperature and the synergist piperonyl butoxide on imidacloprid toxicity to the cat flea (Siphonaptera: Pulicidae). J Econ Entomol 92:1120-1124

Ritzhaupt L, Rowan T, Jones R (1999) Efficacy of selamectin, fipronil, imidacloprid, and lufenuron/milbimycin against *Ctenocephalides felis* in dogs. In: Proc 44th Ann Meet Am Ass Vet Parasitol, New Orleans, Louisiana, July 1999, No. 60, p 57

Robinson WH (1995) Distribution of cat flea larvae in the carpeted household. Vet Dermatol 6:145-150

Rockett CL, Johnston SA (1988) Ectoparasitic arthropods collected from some northern Ohio mammals. Great Lakes Entomol 21:147-149

Role LW (1992) Diversity in primary structure and function of neuronal nicotinic acetylcholine receptor channels. Curr Opin Neurobiol 2:254-262

Rosenheim JA, Johnson MW, Mau RFL, Welter SC, Tabashnik BE (1996) Biochemical preadaptions, founder events, and the evolution of resistance in arthropods. J Econ Entomol 89:263-273

Rosiky B (1978) Health risks associated with animals in different types of urban areas: present status and new ecological conditions due to urbanization. Ann Ist Super Sanita' 14:273-286

Rothschild M (1965) Fleas. Sci Am 213:44-53

Rothschild M (1975) Recent advances in our knowledge of the order Siphonaptera. Ann Rev Entomol 20:241-259

Rothschild M, Clay T (1952) Fleas, flukes and cuckoos. Philosophical Library, New York

Rothschild M, Ford B (1964) Maturation and egg-laying of the rabbit flea (*Spilopsyllus cuniculi* Dale) induced by the external application of hydrocortisone. Nature 203:210-211

Ruf J, Sander E (1990) Bayer internal report, 2 Feb., 1990, AH-D - ID 15904

Rust JH Jr, Cavanaugh DC, O'Shita R, Marshall JD Jr (1971) The role of domestic animals in the epidemiology of plague. I. Experimental infection of dogs and cats. J Infect Dis 124:522-526

Rust MK (1988) An ecological perspective of the host-parasite relationship of the cat flea. In: Proc Nat Urban Entomol Conf, University of Maryland, College Park, Maryland, February 1988, pp 65-71

Rust MK (1992) Influence of photoperiod on egg production of cat fleas (Siphonaptera: Pulicidae) infesting cats. J Med Entomol 29:242-245

Rust MK (1993) Insecticide resistance in fleas. In: Knapp FW (ed) Proc 3rd Int Symp Ectoparas Pets, Lexington, Kentucky, April 1993, pp 18-26

Rust MK (1994) Interhost movement of adult cat fleas (Siphonaptera: Pulicidae). J Med Entomol 31:486-489

Rust MK (1995) Factors affecting control with residual insecticide deposits. In: Rust MK, Owens JM, Reierson DA (eds) Understanding and controlling the German cockroach. Oxford University Press, New York, pp 149-169

Rust MK, Dryden MW (1997) The biology, ecology, and management of the cat flea. Ann Rev Entomol 42:451-473

Rust MK, Reierson DA (1989) Activity of insecticides against the preemerged adult cat flea in the cocoon (Siphonaptera: Pulicidae). J Med Entomol 26:301-305

Rust MK, Hinkle NC, Waggoner M, Mencke N, Hansen O, Vaughn M (2001) The influence of imidacloprid on adult cat flea feeding. Suppl Comp Cont Educ Vet Pract 23 (4A):18-21

Sanderson JP, Roush RT (1995) Management of insecticide resistance in the greenhouse. In: Bishop A, Hausbeck M, Lindquist R (eds) Proc 11th Conf Insect Dis Managem Ornamentals, Fort Myers, Florida, February 1995

Sapre SN (1946) Transmission of pasteurellosis by the fleas (*Ctenocephalides felis*). Ind J Vet Sci 15:151-155

Sargent PB (1993) The diversity of neuronal nicotinic acetylcholine receptors. Ann Rev Neurosci 16:403-443

Sattelle DB (1980) Acetylcholine receptors of insects. Adv Insect Physiol 15:215-315

Sattelle DB, Harrow ID, Hue B, Pelhate M, Gepner JI, Hall LM (1983) Alpha-bungarotoxin blocks excitatory synaptic transmission between cercal sensory neurones and giant interneurone 2 of the cockroach, *Periplaneta americana*. J Exp Biol 107:473-489

Sawicki RM (1979) Resistance of insects to insecticides. Span 22:50-52

Sawicki RM (1987) Definition, detection and documentation of insecticide resistance. In: Ford MG, Holloman DW, Khambay BPS, Sawicki RM (eds) Combating resistance to xenobiotics: biological and chemical approaches. Ellis Horwood, Chichester, England, pp 105-117

Scheidt VJ (1987) Common feline ectoparasites. Part 3. Chigger mites, cat fur mites, ticks, lice, bot fly larvae and fleas. Compan Anim Pract 1:5-15

Scheidt VJ (1988) Flea allergy dermatitis. Vet Clin North Am Small Anim Pract 18:1023-1042

Schein E (1996) Bayer internal report, AH Study No. 112.370

Schelhaas DP, Larson OR (1989) Cold hardiness and wintersurvival in the bird flea, *Ceratophyllus idius*. J Insect Physiol 35:149-153

Schemmer KR, Halliwell REW (1987) Efficacy of alum-precipitated flea antigen for hyposensitization of flea-allergic dogs. Sem Vet Med Surg. 2:195-198

Schick MP, Schick RO (1986) Understanding and implementing safe and effective flea control. J Am Anim Hosp Assoc 22:421-434

Schroeder ME, Flattum RF (1984) The mode of action and neurotoxic properties of the nitromethylene heterocycle insecticides. Pestic Biochem Physiol 22:148-160

Scott DW (1978) Immunologic skin disorders in the dog and cat. Vet Clin North Am Small Anim Pract 8:641

Scott DW (1980) Feline dermatology 1900-1978: A monograph. J Am Anim Hosp Assoc 16:331-459

Sèrtic V (1965) Klinicka i eksperimentalna istrazivanja alergijskih oboljenja u domacih zivotinja. I. Kronicni alergijski dermatitis pasa (Clinical and experimental investigations in allergic diseases of domestic animals. I. Chronic allergic dermatitis of dogs). Veterinarski arhiv, Zagreb 35:278-290

Service MW (1980) A guide to medical entomology. Macmillan Tropical and Subtropical Medical Texts, pp 127-135

Sgard F, Fraser SP, Katkowska MJ, Djamgoz MB, Dunbar SJ, Windass JD (1998) Cloning and functional characterization of two novel nicotinic acetylcholine receptor α subunits from the insect pest *Myzus persicae*. J Neurochem 71:903-912

Shanley K, Douglass PJ, Foil C, Hendrix C, Miller K (1992) Flea and tick control, Part 1. Feline Practice 20:13-22

Shepard HH (ed) (1960) Methods of testing chemicals on insects, Vol. II. Burgess, Minneapolis

Sheppard DC, Joyce JA (1992) High levels of pyrethroid resistance in horn flies (Diptera: Muscidae) selected with cyhalothrin. J Econ Entomol 85:1587-1593

Sheppard DC, Hinkle NC, Hunter III JS, Gaydon DM (1989) Resistance in constant exposure livestock insect control systems: a partial review with some original findings on cryomazine resistance in house flies. Fla Entomol 72:360-369

Shiokawa K, Tsuboi S, Kagabu S, Moriya K (1986) European patent EP 0 192 060 A1.

Shipstone MA, Mason KV (1995) The use of insect development inhibitors as an oral medication for the control of the fleas *Ctenocephalides felis*, *Ct. canis* in the dog and cat. Vet Dermatol 6:131-137

Shmidl JA, Arther RG (1995a) Bayer internal report, 30 March, 1995, AH-D - ID 15359

Shmidl JA, Arther RG (1995b) Bayer internal report, 17 April, 1995, AH-D - ID 15384

Silverman J, Appel AG (1984) The pupal cocoon of the cat flea, *Ctenocephalides felis* (Bouché) (Siphonaptera: Pulicidae): a barrier to ant predation. Proc Entomol Soc Wash 86:660-663

Silverman J, Rust MK (1983) Some abiotic factors affecting the survival of the cat flea *Ctenocephalides felis* (Siphonaptera: Pulicidae). Environ Entomol 12:490-495

Silverman J, Rust MK (1985) Extended longevity of the pre-emerged adult cat flea (Siphonaptera: Pulicidae) and factors stimulating emergence from the pupal cocoon. Ann Entomol Soc Am 78:763-768

Silverman J, Platzer EG, Rust MK (1981a) Infection of the cat flea, *Ctenocephalides felis* (Bouché) by *Neoaplectana carpocapsae* Weiser. J Nemat 14:394-397

Silverman J, Rust MK, Reierson DA (1981b) Influence of temperature and humidity on survival and development on the cat flea, *Ctenocephalides felis* (Siphonaptera: Pulicidae). J Med Entomol 18:78-83

Slacek B, Opdebeeck JP (1993) Reactivity of dogs and cats to feeding fleas and to flea antigens injected intradermally. Aust Vet J 70:313-314

Smit FGAM (1973) Siphonaptera (fleas). In: Smith KGV (ed) Insects and other arthropods of medical importance. British Museum of Natural History, London, pp 325-371

Smith RD, Paul A, Kitron UD, Philip JR, Barnett S, Piel MJ, Ness RW, Evilsizer M (1996) Impact of an orally administered insect growth regulator (lufenuron) on flea infestations of dogs in a controlled simulated home environment. Am J Vet Res 57:502-504

Sobey WR, Menzies W, Conolly D (1974) Myxomatosis: some observations on breeding the European rabbit flea *Spilopsyllus cuniculi* (Dale) in an animal house. J Hyg 72:453-465

Soloway SB, Henry AC, Kollmeyer WD, Padgett WM, Powell JE, Roman SA, Tieman CH, Corey FA, Horne CA (1978) Nitromethylene heterocycles as insecticides. In: Shankland DL, Hollingworth RM, Smyth Jr T (eds) Pesticide and venom neurotoxicity. Plenum, New York, pp 153-158

Soloway SB, Henry AC, Kollmeyer WD, Padgett WM, Powell JE, Roman SA, Tieman CH, Corey RA, Horne CA (1979) Nitromethylene insecticides. In: Geissbühler H, Brooks GT, Kearney PC (eds) Advances in pesticide science part 2. Pergamon Press, Oxford, pp 206-217

Sone S, Nagata K, Tsuboi S, Shono T (1994) Toxic symptoms and neural effect of a new class of insecticide, imidacloprid, on the American cockroach, *Periplaneta americana* (L.). J Pesticide Sci 19:69-72

Sosna CB, Medleau L (1992) The clinical signs and diagnosis of external parasite infestation. Vet Med 78:548-564

Soulsby EJL (ed) (1968) Helminths, arthropods and protozoa of domesticated animals. Baillière, Tindall and Cassell, London

Soulsby EJL (ed) (1982) Helminths, arthropods and protozoa of domesticated animals, 7th edn. Lea & Febiger, Philadelphia

Spande TF, Garraffo HM, Edwards WM, Yeh HJC, Pannell L, Daly JW (1992) Epibatidine: A novel (chloropyridyl)azabicycloheptane with potent analgesic activity from an Ecuadoran poison frog. J Am Chem Soc 114:3475-3478

Stephen S, Rao KNA (1980) Natural occurrence of spotted fever group rickettsiae in the dog flea *Ctenocephalides canis* in Karnataka. Ind J Med Res 71:870-872

Stolper R, Opdebeeck JP (1994) Flea allergy dermatitis diagnosed by intradermal skin tests. Res Vet Sci 57:21-27

Strenger A (1973) Zur Ernährungsbiologie der Larve von *Ctenocephalides felis felis*. B. Zool Jahrb Syst Bd 100:64-80

Stroud RM, Finer-Moore J (1985) Acetylcholine receptor structure, function and evolution. Ann Rev Cell Biol 1:317-351

Studdert VP, Arundel JH (1988) Dermatitis of the pinnae of cats in Australia associated with the European rabbit flea (*Spilopsyllus cuniculi*). Vet Rec 123:624-625

Supperer R, Hinaidy HK (1986) Ein Beitrag zum Parasitenbefall der Hunde und Katzen in Österreich. Dtsch Tierärztl Wochenschr 93:383-386

Tabashnik BE, Finson N, Johnson MW, Moar WJ (1993) Resistance to toxins from *Bacillus thuringiensis* subsp. *kurstaki* causes minimal cross-resistance to *B. thuringiensis* subsp. *aizawai* in the diamondback moth (Lepidoptera: Plutellidae). Appl Environ Microbiol 59:1332-1335

Taryannikov VI (1983) Parasites of the jackal *Canis aureus aureus* L. in the middle reaches of the Syr-Dar'ya River. Parazitologiya 17:478-480

Tas (1990) Exposure 1. Chronic dietary exposure analysis. In: Petersen SR (ed) Technical assessment systems Inc., Washington, DC

Tattersfield F, Morris HM (1924) An apparatus for testing the toxic values of contact insecticides under controlled conditions. Bull Ent Res 14:223-233

Thomas RE, Wallenfels L, Popiel I (1996) On-host viability and fecundity of *Ctenocephalides felis* (Siphonaptera: Pulicidae), using a novel chambered flea technique. J Med Entomol 33:250-256

Thornhill R, Alcock J (eds) (1983) Timing of mate locating. In: Thornhill R, Alcock J (eds) The evolution of insect mating systems. Harvard University Press, Cambridge, pp 90-118

Thyssen J, Machemer L (1999) Imidacloprid: Toxicology and metabolism. In: Yamamoto I, Casida JE (eds): Nicotinoid insecticides and the nicotinic receptor. Springer-Verlag, Tokyo, pp 213 - 222

Timm KI (1988) Pruritus in rabbits, rodents and ferrets. Vet Clin N Am Small Anim Pract 18:1077-1091

Toll PA (1990a) Mobay Corp. internal report, 1 March, 1990, No. 100059

Toll PA (1990b) Mobay Corp. internal report, 22 Aug., 1990, No. 100238

Toll PA (1990c) Mobay Corp. internal report, 30 Aug., 1990, No. 100241

Tomizawa M, Yamamoto I (1993) Structure-activity relationships of nicotinoids and imidacloprid analogs. J Pesticide Sci 18:91-98

Tomizawa M, Otsuka H, Miyamoto T, Yamamoto I (1995a) Pharmacological effects of imidacloprid and its related compounds on the nicotinic acetylcholine receptor with its ion channel from the *Torpedo* electric organ. J Pesticide Sci 20:49-56

Tomizawa M, Otsuka H, Miyamoto T, Eldefrawi ME, and Yamamoto I (1995b) Pharmacological characteristics of insect nicotinic acetylcholine receptor with its ion channel and the comparison of the effect of nicotinoids and neonicotinoids. J Pesticide Sci 20:57-64

Torgeson P, Breathnach R (1996) Flea dermatitis and flea hypersensitivity: The current situation in Ireland. Irish Vet J 49:426-433

Tovar RM (1947) Infection and transmission of *Brucella* by ectoparasites. Am J Vet Res 8:138-140

Tränkle SB (1989) Wirtsspezifität und Wanderaktivität des Katzenflohes *Ctenocephalides felis* (Bouché). Diplomarbeit, Universität Freiburg, Freiburg

Traub R, Wisseman Jr CL, Farhang-Azad A (1978) The ecology of murine typhus - a critical review. Trop Dis Bull 75:237-317

Tsunoyama K, Gojobori T (1998) Evolution of nicotinic acetylcholine receptor subunits. Mol Biol Evol 15:518-527

Urquhart GM, Armour J, Duncan J, Dunn AM, Jennings FW (eds) (1987) Order Siphonaptera. In: Veterinary parasitology. Longman Scientific & Technical, Essex, England, pp 171-175

Van Den Beukel I, Van Kleef RGDM, Zwart R, Oortgiesen M (1998) Physostigmine and acetylcholine differentially activate nicotinic receptor subpopulations in *Locusta migratoria* neurones. Brain Res 789:263-273

Van Winkle KA (1981) An evaluation of flea antigens used in intradermal skin testing for flea allergy in the canine. J Am Anim Hosp Assoc 17:343-354

Vaughan JA, Coombs ME (1979) Laboratory breeding of the European rabbit flea, *Spilopsyllus cuniculi* (Dale). J Hyg 83:521-530

Veith I (1989) Topical therapy for flea-allergic pets. Vet Med 84:588-592

Venard CE (1938) Morphology, bionomics and taxonomy of the cestode *Dipylidium caninum*. Ann NY Acad Sci 37:273-328

Venter JC, Di Portio U, Robinson AD, Shreeve SM, Lai J, Kerlavage AR, Fracek Jr SP, Lentes KU, Fraser CM (1988) Evolution of neurotransmitter receptor systems. Prog Neurobiol 30:105-169

Wade SE, Georgi JR (1988) Survival and reproduction of artificially fed cat fleas *Ctenocephalides felis* (Bouché) (Siphonaptera: Pulicidae). J Med Entomol 25:186-190

Wege PJ (1994) Challenges in producing resistance management strategies for *Myzus persicae*. In: Proc Brighton Crop Prot Conf - Pests and Diseases, pp 419-425

Wen Z, Scott JG (1997) Cross-resistance to imidacloprid in strains of German cockroach (*Blatella germanica*) and house fly (*Musca domestica*). Pestic Sci 49:367-371

Werner G, Hopkins T, Shmidl JA, Watanabe M, Krieger K (1995a) Imidacloprid, a novel compound of the chloronicotinyl group with outstanding insecticidal activity in the on-animal treatment of pests. Pharmacol Res 31 (Suppl):136

Werner G, Hopkins T, Shmidl JA, Watanabe M, Krieger K (1995b) Bayer internal report, 19 June, 1995, AH-D - ID 15423

Werner SB, Weidmer CE, Nelson BC, Nygaard GS, Goethals RM, Poland JD (1984) Primary plague pneumonia contracted from a domestic cat at South Lake Tahoe, California. J Am Med Assoc 251:929-931

Wheeler CM, Douglas JR (1941) Transmission studies of sylvatic plague. Proc Soc Exp Biol Med 47:65-66

White SD, Ihrke PF (1983) Allergic diseases. In: Pratt PW (ed) Feline medicine. Am Vet Publications, Santa Barbara, pp 577-580

Whiting P, Lindstrom J (1987) Purification and characterization of nicotinic acetylcholine receptor from rat brain. Proc Natl Acad Sci USA 84:595-599

Williams B (1983) The cat flea, *Ctenocephalides felis* (Bouché): its breeding biology, and its larval anatomy compared with that of two Ceratophylloid larvae. Ph.D. Dissertation, University of Oxford, Oxford

Williams B (1986) One jump ahead of the flea. New Sci 31:37-39

Williams B (1993) Reproductive success of cat fleas, *Ctenocephalides felis*, on calves as unusual hosts. Med Vet Entomol 7:94-98

Williams SG, Sacci JB, Schriefer ME, Anderson EM, Fujioka KK, Sorvillo FJ, Barr AR, Azad AF (1992) Typhus and typhus-like rickettsiae associated with opossums and their fleas in Los Angeles County, California. J Clin Micro 30:1758-1762

Woodford JAT, Mann JA (1992) Systemic effects of imidacloprid on aphid feeding behaviour and virus transmission on potatoes. In: Proc Brighton Crop Protect Conf - Pests and Diseases, pp 557-562

World Health Organisation (WHO) (1988) Urban vector and pest control. Tech Rep Ser No. 767, WHO, Geneva

World Health Organization (WHO) (1992) Present status of pesticide resistance. Tech Rep Ser No. 818, WHO, Geneva, pp 2-17

World Health Organisation (WHO)/World Small Animal Veterinary Association (WSAVA) (1981) Guidelines to reduce human health risks associated with animals in urban areas. WHO, Geneva

Yamamoto I (1965) Nicotinoids as insecticides. Adv Pest Control Res 6:231-260

Yamamoto I (1996) Neonicotinoids - Mode of action and selectivity. Agrochem Japan 68:14-15

Yamamoto I, Tomizawa M, Saito T, Miyamoto T, Walcott EC, Sumikawa K (1998) Structural factors contributing to insecticidal and selective actions of neonicotinoids. Arch Insect Biochem Physiol 37:24-32

Yamamoto I, Yabuta G, Tomizawa M, Saito T, Miyamoto T, Kagabu S (1995) Molecular mechanism for selective toxicity of nicotinoids and neonicotinoids. J Pesticide Sci 20:33-40

Yasuda F, Hashiguchi J, Nishikawa H, Watanabe S (1968) Studies on the life-history of *Dipylidium caninum* (Linnaeus 1758). Bull Nippon Vet Zootech Coll 17:27-32

Yeruham I, Hadani A, Sklar A (1982) Infestation of calves with cat flea *Ctenocephalides felis felis* (Bouché 1835). Ref Vet 39:8-10

Yeruham I, Rosen S, Hadani A (1989) Mortality in calves, lambs and kids caused by severe infestation with the cat flea *Ctenocephalides felis felis* (Bouché 1835) in Israel. Vet Parasitol 30:351-356

Young D (1995) Bayer internal report, 15 Dec., 1995, AH-D - ID 15793

Young DR, Ryan WG (1999) Comparison of Frontline® Top Spot™, Preventic® collar alone or combined with Advantage® in control of flea and tick infestations in water immersed dogs. In: Proc 5th Int Symp Ectoparas Pets, Fort Collins, Colorado, April 1999, No. 24

Young DR, Young R, Arther RG (1996) Efficacy of Advantage™ (9.1% imidacloprid) for control of fleas on cats and large dogs. In: Proc 41st Ann Meet Am Ass Vet Parasitol, Louisville, Kentucky, July 1996, No. 58, p 49

Yutuc LM (1968) The cat flea hitherto unknown to sustain the larva of *Dipylidium caninum* (Linnaeus 1758) from the Philippines. Philipp J Sci 97:285-289

Zajicek D (1987) Laboratory diagnosis of parasites in the Czech Socialist Republic in the period 1976-1986. IV. Dogs, cats. Veterinarstvi 37:549-550

Zakson M, Gregory LM, Endris RG, Shoop WL (1995) Effect of combing time on cat fleas (*Ctenocephalides felis*) recovery from dogs. Vet Parasitol 60:149-153

Zakson-Aiken M, Gregory LM, Shoop WL (1996) Reproductive strategies of the cat flea (Siphonaptera: Pulicidae): Parthenogenesis and autogeny? J Med Entomol 33:395-397

Zielhuis RL, Van Den Kerk FW (1978) The use of a safety factor in setting health based permissible levels for occupational exposure. Int Arch Occup Hlth 42:191-201

Zimmerli J (1982) Study of the parasites of the stone marten (*Martes foina*) in the Canton of Vaud during the period 1980-1981. Schweiz Arch Tierheilkd 124:419-422

Zlotorzycka J, Eichler W, Ludwig HW (eds) (1974) Taxonomie und Biologie der Mallophagen und Läuse mitteleuropäischer Haus- und Nutztiere. Parasitologische Schriftenreihe, Bd 22, VEB Verlag Gustav Fischer, Jena

Zwart R, Oortgiesen M, Vijverberg HPM (1994) Nitromethylene heterocycles: Selective agonists of nicotinic receptors in locust neurons compared to mouse N1E-115 and BC3H1cells. Pestic Biochem Physiol 48:202-213

# Subject Index

Abott formula   107
acanthosis   42, 46 - 47
accumulation   82, 83, 146
acetylcholine receptor
– muscarinic (mAChR)   69
– nicotinie (nAChRs)   66, 67, 69, 70
– – insecticidal   70, 82
– – mammalian   70
– – subunit composition   70
acetylcholinesterase   61-62
active ingredient (a.i.)   64, 83, 88-89, 144
adult   1, 17, 18, 26, 27-34, 142-143
– environmental conditions   30
– feeding   18, 23, 29-34
– fertilization
– longevity   31, 33
– preemerged (see there)   17, 27-29
– prexisting   136, 139
adulticide   24, 104, 126, 128
agrochemical   68, 78-79, 92, 94, 105
allergy / allergic   39-44
– flea allergy dermatitis (see FAD)   13, 39, 41-56
– hypersensitivity (see flea bite
  hypersensitivity)   39
– stages   43, 47, 49
alopecia (hair loss)   40-44, 114-115, 127, 152,
  157, 161
α-BGT (α-bungarotoxin)   70-71
α subunit   70, 74, 82
amitraz   136, 151
anemia, iron deficiency   13, 40, 43, 52
animal size   135
antibiotics   46, 55, 161
antibody   47, 56
anticoagulant   32, 47, 49
antifeedant effect   76-77
antigenic components   13, 40, 44, 46-49, 51-
  53, 56, 77
antihistamines   53, 55
aphid
– green peach (Myzus persicae)   65, 70, 73-74,
  76-77, 79, 82
– tobacco (Myzus nicotianae)   76-79
Apis mellifera (honey-bee)   73
application   95, 99, 100
– of imidacloprid   92, 95, 99, 139-140
– interval   95, 109, 144
– site   92, , 99, 101-102, 112, 153
– volume   92, 100, 101, 109
Archaeopsylla erinacei (hedgehog flea)   1, 3,
  4, 7, 32, 51
area / site
– application site   99

– binding site of imidacloprid   67, 70, 72-74
– – vertebrate, imidacloprid   70, 71, 73-75, 82
– breeding sites   36
– resting areas   143, 146
– sleeping areas   143
– target site resistance   6
arrow-poison frog (Epipedobates tricolor)   63
Arthropoda   1
atopy   41, 44, 46-47, 50
Australia   9, 107-108, 111

bacterial dermatitis / dermatose, secondary
– purulent   40
– pyoderma   45, 55, 152
bath   135, 152
bedding   151
behavior
– alteration   76-77
– host-seeking   28, 29
Bemisia
– argentifolii (silverleaf whitefly)   79
– tabaci (cotton whitefly / tobacco whitefly)
  65, 77
binding site of imidacloprid   67, 70, 72-74
biological flea control   58
biopsy   46, 52
biphasic effect   72
bitch
– lactating   89 - 90, 118
– pregnant   90, 118
biting resp. chewing lice (Trichodectes canis)
  160-161
blanket   137, 144-145
Blatella germanica (German cockroach)   73,
  81-82
blood, concentration of imidacloprid   101-103
body surface   95, 101, 106
breeding sites   36
breeds   110
Brucella
– abortus   14
– melitensis   14
– suis   14
α-bungarotoxin (α-BGT)   70-71

carpet   21-22, 24-26, 28, 30, 36, 57-58, 62, 124, 138
carrier, compound   101, 104
cat flea (Ctenocephalides felis)   1, 3, 4-7, 14,
  17-34, 59-62, 78, 81, 154-155, 157
CBH (cutaneous basophilic hypersensitivity)
  49-50

*Cediopsylla simplex* (common eastern rabbit flea) 7, 154
Ceratophyllidae 1
*Ceratophyllus (C.)*
– *columbae* 1
– *gallinae* (poultry flea) 3, 4, 7
chewing resp. biting lice (*Trichodectes canis*) 160-161
*Cheyletiella parasitivorax* 14
chitin 136
chlordane 59
chloronicotinyl (neonicotinoides) 63, 65, 68, 70, 73-75, 79
chlorpyrifos 60, 78
coat (*see* hair) 95, 101, 110, 135, 137-138, 142, 147, 153
cockroach
– American (*Periplaneta americana*) 71-73, 81-82
– German (*Blattella germanica*) 73, 81-82
cocoon 25-29, 35, 37-38, 58, 62, 154, 161
Coleoptera 68
combs (*see* ctenidia) 3, 6-7
conduction
– block 71-72
– multi-conductance state 72
– nerve conduction 72
control of flea
– biological flea control 58
– efficacy studies 106-107, 109, 112, 117, 126, 136, 138, 143, 146, 153, 157
– mouse 159
– persistency of 100
– program 147
– rat 159
control of lice 160-161
control of sheep ked 161-162
corticosteroids 53, 55, 161
cotton whitefly / tobacco whitefly (*Bemisia tabaci*) 65, 77, 79
cottontail, flea strain 80, 136-137
counting 107, 108
– comb counting 107
– technique 108
– thumb / manual counting 107
coxal depressor motorneuron $D_f$ 71, 82
ctenidia 3, 6-7
– genal comb 3, 6-7
– pronatal comb 3, 6-7
*Ctenocephalides*
– *canis* (dog flea) 1, 3-7
– *felis* (cat flea) 1, 3-7
– *felis damarensis* 5
– *felis orientis* 5
– *felis strongylus* 5

cutaneous basophilic hypersensitivity (CBH) 49-50
cysticercoids 14
cytochrome P450 61, 87

debris 21-22, 25, 51, 57-56, 141-143, 145-146
degeneration 72, 132-133
*Dermacentor variabilis* (American dog tick) 151
dermatitis / dermatose
– acute serous (wet eczema) 42, 47
– flea allergy (*see* FAD) 13, 39, 40–56
– flea bite 39
– miliary (*see* feline miliary dermatitis / feline exzema) 13, 40, 42-43, 46, 53
– purulent (*see* secondary bacterial dermatose / dermatitis) 40-42, 45, 118
– pyoderma (*see* secondary bacterial dermatose / dermatitis) 45, 55, 152
desensitization therapy (*see* hyposensitization) 50, 55
detoxification 61, 79-81
*Diamanus montanus* (ground squirrel flea) 7
diapause 14, 35,
diazinon 60, 78, 135, 157
dieldrin 59, 78
*Dipetalonema reconditum* 13, 14
Diptera 1
*Dipylidium caninum* (dog tapeworm) 13-14, 43, 52
*Dirofilaria immitis* (heartworm) 13, 14
disease transmission (*see* transmission of disease) 13-15, 161
distribution
– of fleas
– – on hosts 35-36
– – worldwide 3, 9-12
– of imidacloprid 81, 101-106
dog
– flea (*Ctenocephalides canis*) 1, 3-7
– seeder dog 137
– tapeworm (*Dipylidium caninum*) 13-14, 43, 52
dose / dosage
– confirmation 109-111, 120
– response 109, 111
– therapeutic 110–112
– titration 109
*Drosophila melanogaster* (fruit fly) 73

*Echidnophaga gallinacea* (sticktight flea)   3, 7
ecotoxicity   92–94
– aquatic   92
– terrestrial   92
– wildlife   92
eczema
– feline (feline miliary dermatitis)   13, 40,
   42-43, 46, 53
– summer   40
– wet (acute serous dermatitis)   42, 47
efficacy   107
– clinical efficacy in cats and dogs   106–127
– controlled efficacy studies   106-113
– duration of   108-127
– enhancement of   148
– evaluation of   107
– flea   106-127, 154-19
– onset of   109
egg   17-20
– deposition   18-19
– output   18
– production   18-19
electron microscopy   133, 134
electron-withdrawing group   74
embryotoxicity   87
emergence, stimuli   26-35
environment / environmental conditions
   149–152
– adult / premerged adult   27, 29-30
– hatching   19
– larva   36, 141
– pupa   26, 38
– study, simulated home environment   123-
   124, 137-138, 144
– treatment of   140
eosinophils   47, 52
epibatidine   63, 65
epidemiology   15, 33, 35–38, 139
epidermal scales   142, 146-147
*Epipedobates tricolor* (arrow-poison frog)   63
epipharynx   32
erythema   41, 44, 48, 52, 55, 88
Europe / European   157-159
ferret (*Mustela putorius furo*)   157–159
– mouse flea (*Leptopsylla segnis*)   7
– rabbit
– – (*Oryctolagus cuniculus*)   154
– – (*Spilopsyllus cuniculi*)   7, 154
evaluation of efficacy   107
exposure, human oral   95-98

FAD (flea allergy dermatitis)   13, 39, 40–56
– clinical signs   41
– diagnosis   51, 52
– – antigen-specific IgE   52

– – biopsy   46, 52
– – eosinophils   46, 52
– – flea combing   51
– – intradermal skin testing (IDST)   52-53
– – skin scrapings   51
– differential diagnosis   46
– immunopathogenesis   47
– localisation   42, 43
– phases
– – in cats   50
– – in dogs   49
– – in guinea pigs   44, 48
– resolution of   115
– self trauma   41
– treatment of   54–56
feline eczema (feline miliary dermatitis)   13,
   40, 42-43, 46, 53
fenthion   60, 89, 138
ferret, European (*Mustela putorius furo*)
   157-159
– efficacy, flea   158
– tolerability   158
fipronil   61, 77, 80, 126-128, 131, 135-136, 140,
   143, 150-151, 161
flea
– allergy dermatitis (*see* FAD)   13, 39, 40–56,
   115
– blood consumption   13, 39
– bite
– – dermatitis   39-40
– – hypersensitivity   39, 41, 49, 157
– burden   50, 115, 117, 124-125, 135, 138, 140,
   156
– combing   51, 107-108, 124-125, 138, 158
– count (*see* counting)   107, 108
– efficacy, rabbit flea   154-157
– feces   21-24, 32, 41, 61, 142
– saliva   13, 32, 40, 44, 47, 49-51, 76-77
– speed of kill   108, 128-131
– strains   23, 59-61, 80, 107, 113, 120, 135-136
– survival   30
– winter fleas   37
Florida   10, 36, 61, 139-140
folliculitis   41-42, 46
France   9, 114, 126
friend leucemia virus (FLV)   14
fruit fly (*Drosophila melanogaster*)   73
fur (*see* hair)   95-101, 110, 135, 137-138, 142,
   147, 153

GABA (γ-aminobutyric acid)   69
genotoxicity   87
geotaxis   21, 30
Germany   9-11, 109, 117, 155
glutathione transferase   61

green
- peach aphid (*Myzus persicae*) 65, 70, 73-74, 76-77, 79, 82
- rice leafhopper (*Nephotettix cincticeps*) 65
grooming 32

hair (coat) 95-101, 110, 135, 137-138, 142, 147, 153
- alopecia (hair loss) 40-44, 114-115, 127, 152, 157, 161
- coat length / coat type 110, 142
- concentration of imidacloprid on 101-104
hatching 19
- environmental conditions 19-20
heartworm (*Dirofilaria immitis*) 13, 14
hedgehog flea (*Archaeopsylla erinacei*) 1, 3, 4, 7, 32, 51
hemoglobin 23
hemorrhage 40
hexachlorocyclohexane (HCH) 59, 136
histamine 47, 49, 53, 55
honey-bee (*Apis mellifera*) 73
honeydew excretion 76-77
host
- distribution on hosts 35-36
- interhost movement 33
- parasite-host-relationship 33
- spectrum 36
host-seeking behavior 28, 29
housefly (*Musca domestica*) 73
household 125, 126
- multi-pet / multiple animal 125, 126
- single animal 126
human flea (*Pulex irritants*) 3, 7
hygrotaxis 21
*Hymenolepis* 14
- *citelli* 14
- *diminuta* 14
- *microstoma* 14
- *nana* 14
hyperkeratosis 42, 45
hyperpigmentation 42
hypersensitivity 13
- allergic 39
- flea bite 39
hyposensitization (desensitization therapy) 50, 55

IDI (insect development inhibitor) 136
IDST (intradermal skin testing) 52-53
IgE, antigen-specific 52
IGR (insect growth regulator) 136

imidacloprid
- agonistic effects 67, 69, 71-72
- analytic extraction 144, 145
- antagonistic effects 71-73
- application 77, 79, 82, 86, 90, 92, 95, 98-99, 100, 102, 108
- binding site 67, 70, 72-74
- biological profile 67, 68
- chemical properties 64-67
- clinical efficacy in cats and dogs (*see also* efficacy) 106–127
- concentration 101–103
- – blood 103
- – hair 101-102
- – skin 102-103
- depolarization 69, 72
- distribution 81, 101-106
- environmental and habitual factors 149–152
- history 63
- hydrophobicity 66
- insecticidal
- – activity 67-75
- – effects, indirect 76–78
- larvicidal effect 141-147
- metabolism 81, 83, 84
- mode of action 68–75
- penetration 66-67, 77, 81
- physiochemical properties 66
- resistance of sucking pests against imidacloprid 78–81
- safety (see there)
- selectivity 74, 85
- sunlight degradation 93
- tolerability 88–90
- toxicology and pharmacology 81–94
- trace amound 138
- vertebrate 73
- – binding sites 82
- water degradation 93
imidazolidine ring 74
infestation
- level 115
- louse 161
inhibition 142
insect / Insecta 1
- IDI (insect development inhibitor) 136
- IGR (insect growth regulator) 136
- reproduction 76
insecticides / insecticidal 26
- concentration 76, 101
- tolerance 62
interhost movement 33
IPM (integrated pest management) 80
IRM (integrated resistance management) 80
ivermectin 136

keeping conditions   107, 135
kill, speed of flea kill   108, 128-131
kitten   13, 43, 89-92, 99, 153-154,
knock-down effect   63, 109-110

label recommendation   99
lactating bitch   89-90, 118
*Laodelphax striatellus* (small brown
  planthopper)   79
larva   17
– development   20-24
– environmental conditions   23-24
– instar   17, 20-22, 24, 141
– microhabitat   21
– mortality   23
– nutrition   21
larvicidal   137
– activity   124
– effect of imicacloprid   141-147
$LD_{50}$   81, 83
Lepidoptera   68
*Leptopsylla segnis* (European mouse flea)   7
leucemia, friend leucemia virus (FLV)   14
lichenification   42
life cycle   17
light
– microscopy   132
– traps   30, 57-58
*Linognathus setosus* (sucking lice)   160-161
litter, unweaned   153
louse / lice
– biting resp. chewing (*Trichodectes canis*)
  160
– control   161
– infestation   161
– sucking (*Linognathus setosus*)   160
lufenuron   61, 89, 136, 138, 140-141

malathion   60, 78
market   12
– agrochemical market   68
– animal health market   12
– companion animal market   12
– ectoparasiticide market   12
*Melophagus ovinus* (sheep ked)   161, 162
membranes, intersegmental   132
metamorphosis   17
methoprene   136
microscopy   132–134
– electron   133, 134
– light   132
milbemycin   89, 131, 135
miliary dermatitis (*see* feline miliary
  dermatitis)   13, 40, 42-43, 46, 53

morphology   3, 4, 6
mouse flea
– control of   159
– European (*Leptopsylla segnis*)   7
movement
– interhost   33
– patterns   143
multi-pet / multiple animal   117, 125, 126
*Musca domestica* (housefly)   73
*Mustela putorius furo* (European ferret)   157
mutagenicity   87
*Myzus*
– *nicotianae* (tobacco aphid)   76-79
– *persicae* (green peach aphid)   65, 70, 73-74,
  76-77, 79, 82

neonicotinoides (chloronicotinyl)   63, 65, 68,
  70, 73-75, 79
*Nephotettix cincticeps* (green rice leafhopper)
  65
nerve conduction   72
neuron   68
neurotransmitter   69
nicotine   63, 66, 70, 73-76, 79, 82-83
– acetylcholine receptors (nAChRs)   66-67,
  69-70
nithiazine   63
nitromethylene heterocycles   63
NOEL (no-observed-effect level)   85-87
*Nosopsyllus fasciatus* (northern rat flea)   7

*Odontopsyllus multispinus* (giant eastern
  rabbit flea)   154
off-host stages   62, 136, 145, 147
*Orchopeas howardii* (squirrel flea)   7
*Oryctolagus cuniculus* (European rabbit)
  154
ovicidal   137-138
oviposition   19
oxidase, microsomal   81

papules   40-41
parakeratosis   45-46
parasite-host-relationship   33
parental investment   23
*Pasteurella sp.*   14
PBO (piperonyl butoxide)   78, 79, 148
*Periplaneta americana* (American
  cockroach)   71-73, 81-82
permethrin   60, 77-78, 135
persistency of control   100
pest
– management strategies   62

– – IPM (integrated pest management) 66, 76, 80
– – IRM (integrated resistance management) 80
– resistance of sucking pests 61, 78–81
pet 12-14, 28, 140
– multi-pet / multiple animal household 117, 125, 126
– owner 12-14, 28, 35, 41, 51, 54, 89, 95, 98, 127, 143
– rabbit 154–157
phaeochromocytoma (PC12) cells 72
phallosome 5
pharmacokinetic 83, 84, 109
– absorption 83
– accumulation 83
– elimination 83
– excretion 83
photostability 64
phototaxis 21, 30
piperonyl butoxide (PBO) 78, 79, 148
plague (*Yersinia pestis*) 14-15
planthopper, small brown (*Laodelphax striatellus*) 79
poultry flea (*Ceratophyllus gallinae*) 3, 4, 7
predisposition 50
preemerged adult 27-29
– environmental conditions 27
– longevity 27
pregnant bitch 90, 118
premise treatment
– indoor 56-58
– outdoor 58
prepupa 25-28, 58
prevalence 9-11, 114
proboscis 32, 49
protandry 31
protection 21, 26, 28, 62, 123, 136, 138, 153, 156, 162
protogony 31
pruritus (scratching) 39-45, 48, 52, 55, 114, 118
public health 12
*Pulex irritants* (human flea) 3
Pulicidae 1
pupa / pupal 17, 20, 23, 25–28, 35, 142, 161
– environmental conditions 26, 30, 38, 58
– prepupa 25, 58
– pupal window 28
pupation 25
puppy 13, 43, 89, 91, 99, 118, 153
purulent dermatitis (*see* secondary bacterial dermatitis) 40-42, 45, 118
pustules 40
pyoderma (secondary bacterial dermatitis) 45, 55, 152
pyriproxyfen 136, 138

queen
– gravid 89, 92
– lactating 89, 92, 153-154

rabbit 154-157
– European (*Oryctolagus cuniculus*) 154
– North America cotton tailed (*Sylvilagus floridanus*) 154
rabbit flea
– common eastern (*Cediopsylla simplex*) 7, 154
– control of 155-157
– European (*Spilopsyllus cuniculi*) 1, 7, 19, 51, 154
– flea development 154
– giant eastern (*Odontopsyllus multispinus*) 154
rainfall 94, 149-150
rat flea
– northern (*Nosopsyllus fasciatus*) 7
– oriental (*Xenopsylla cheopis*) 7, 15
receptor 66-67, 69-75, 82
recommendation / label recommendation 96, 99, 118, 122, 127-128, 139, 153
reinfestation
– artificial 106-110, 112, 120-123, 128-131, 135, 141, 149-151, 155-156, 158
– natural 36-38, 62, 98, 112-113, 115-118, 125, 143, 147, 159
repellency 76, 94, 135
reproduction 18-19, 31-32, 35, 76
reservoir 36, 136
residual activity 56, 61, 64, 109-110, 112, 115-120, 125-126, 138-140, 149, 157-158, 161
resistance 59–62, 78-81
– baseline 62, 81
– bioassays 59, 77, 79
– cross- 61, 80
– definition of 59, 78
– genes 59, 61, 79-80
– genetic mechanism 59, 61, 78
– IRM (integrated resistance management) 80
– metabolic degradation 67, 78-79
– monitoring 62, 80-81
– ratio 59-60, 79
– target site 61, 78-79
resting areas 58, 143, 146-147, 161
*Reticulitermes flavipes* (termites) 68, 77
*Rhipicephalus sanguineus* (brown dog tick) 128, 151
*Rickettsia thyphi* (murine typhus) 14-15
risk assessment 85, 92, 94-98
roaming activity 124, 138

rodents, control of flea on 159
- mouse 159
- rat 159
rotenone 136

safety
- in man 94-98
- margin of 83, 89, 93, 95-98
- target animal 85-87, 88-92, 153
sclerotization 23
scratching (pruritus) 39-43, 45, 48, 52, 55,
  114, 118
seborrhea 42-43, 45
secondary bacterial dermatitis / dermatose
- purulent 40-42, 45, 118
- pyoderma 45, 55, 152
seeder dog 137, 142
selamectin 128-135, 143
shampoos 55, 149-150, 152
sheep ked (*Melophagus ovinus*) 161-162
- control of 161-162
single-channel currents 72
siphonaptera 1
site (*see* area)
skin
- concentration of imidacloprid 102-103
- IDST (intradermal skin testing) 46, 52-53
- scrapings 46, 51
sleeping areas 58, 143, 146-147
sperm precedence 31
spermatheca 3, 31
*Spilopsyllus cuniculi* (European rabbit flea)
  1, 7, 19, 51, 154
spines (*see* ctenidia) 3, 6-7
spot-on formulation 99-101, 104-105, 128
spreading (distribution) 99, 100-106
squirrel flea (*Orchopeas howardii*) 7
- ground squirrel flea (*Diamanus montanus*)
  7
steam cleaning 57
sticktight flea (*Echidnophaga gallinacea*) 3, 7
stroking test 95-98
study
- comparative field 136–141
- comparative laboratory 127-136
- controlled efficacy 107, 110-113, 120-121
- dose / dosage
- - confirmation 110-113
- - response 108-109
- - titration 109
- field 80, 94, 107, 113-120, 125-127, 136-141
- laboratory 76-77, 93-94, 107, 108-113, 120-
  125, 155-157, 157-159
- simulated home environment 123-124,
  137-138, 144-146

subconductance currents 72
α subunit 70, 74, 82
sucking
- insects 68
- lice (*Linognathus setosus*) 160-161
- pests, resistance against imidacloprid 78-
  81
surroundings 35, 56-58, 141
survival 19-20, 23-24, 26-27, 30-31, 34, 37-38
- indoor 24, 38
- outdoor 23-24, 27-28, 37-38
susceptibility 59, 61-62, 79
swimming 94, 115, 149-150
*Sylvilagus floridanus* (North America cotton
  tailed rabbit) 154
synergism 79, 148

target
- animal safety 85-92, 94
- site resistance 61, 78
taxonomy 1, 3, 5-7
teratogenicity 87-88
termites (*Reticulitermes flavipes*) 68, 77
thigmotaxis 21
tick 127-128, 160
- American dog (*Dermacentor variabilis*)
  151
- brown dog (*Rhipicephalus sanguineus*)
  128, 151
tobacco
- aphid (*Myzus nicotianae*) 76-79
- whitefly / cotton whitefly (*Bemisia tabaci*)
  65, 77, 79-80
tolerability of imidacloprid 88-92
- cats 90-92
- dogs 90-91
- ferrets 158
- ingestion 89
- rabbit 156-157
*Torpedo* electric organ 71, 73, 82
toxicity
- birds 93-94
- ecotoxicity (*see there*) 92-94
- fish 83, 93
- genotoxicity 87
- insect 65-66, 68, 71-72, 76-78, 81-83, 93
- mammalian 73-75, 82-83, 85-94, 156-157
- repeated-dose, chronic 85-87, 97, 156
- - dermal 85-86
- - inhalation 86-87
- - oral 85-87
- reproductive 87
- single-dose, acute 85-86, 97
- - dermal 85-86
- - eye irritation 85, 88

– – inhalation -8685
– – oral 85-86
– symptoms
– – fleas 77-78, 82, 105, 131-134
– –insects in general 69, 81-82
– – sheep ked 161
– temperature effect 106, 148
transmission of disease 13-15, 161
treatment
– concomitant 55, 115, 126-127, 135, 140
– desensitization therapy (*see*
   hyposensitization) 55-56
– environmental 54, 56-58, 140-147
– FAD (flea allergy dermatitis) 54–56
– on-host treatment 99-101, 136-138
– premise treatment (*see there*) 24, 56–58
*Trichodectes canis* (biting lice) 160-161
triflumuron 136
typhus, murine (*Rickettsia thyphi*) 14-15

umbrella effect 153, 154
USA 3, 10-13, 15, 40, 59107, 157

vacuolization 72, 134
vacuuming 25, 51, 57, 58, 147, 151
volume of application 100-101, 108-109

washing 57, 151
water
– degradation of imidacloprid in 93
– exposure 149
– immersion 149-152
welfare combing 124, 138
white sock technique 142
whitefly
– cotton / tobacco (*Bemisia tabaci*) 65, 77,
   79-80
– silverleaf (*Bemisia argentifolii*) 79
winter fleas 37-38
worldwide distribution of fleas 3, 5, 9-11

*Xenopsylla cheopis* (oriental rat flea) 7, 15

*Yersinia pestis* (plague) 14-15